Topics in Applied Physics Volume 1

Founded by Helmut K. V. Lotsch

Dye Lasers

Edited by F. P. Schäfer

With Contributions by
K. H. Drexhage · T. W. Hänsch · E. P. Ippen
F. P. Schäfer · C. V. Shank · B. B. Snavely

With 114 Figures

Springer-Verlag Berlin · Heidelberg · New York 1973

Professor Dr. FRITZ PETER SCHÄFER

Max-Planck-Institut für Biophysikalische Chemie, D-3400 Göttingen

ISBN 3-540-06438-9 Springer-Verlag Berlin-Heidelberg-New York
ISBN 0-387-06438-9 Springer-Verlag New York-Heidelberg-Berlin

Introduction to the Series «Topics in Applied Physics»

Recent progress in pure and applied research has produced vast quantities of results. The great need that this creates for authoritative reviews cannot be met by the scientific journals. Textbooks are, by their very nature, unsuitable for the discussion of specific topics in depth, and as a consequence the methods and results of newer research tend to receive rather cursory treatment. Advanced monographs usually appeal exclusively to experts in a particular field, and individuals find it expensive to purchase symposium proceedings or a collection of review papers for the sake of a particular article when the rest of the volume is of marginal interest.

«Topics in Applied Physics» is a new series, published by Springer-Verlag and devoted to *critical reviews of subjects of current interest in applied physics*. This "monograph" series is intended to fill the gap described above. Each volume deals with a particular topic, and contributions are invited by an editor who is a recognized authority in the field in question. The authors are scientists who are actively engaged in advancing the frontiers of research and thus write with the authority that comes from personal involvement.

This series is designed to provide the necessary background, theory, and working information on particular topics for physicists (and chemists), engineers, and advanced students. Furthermore, these critical reviews are definitive and intensive enough to be used by a specialist in some aspect of the topic wishing to update his knowledge on related areas. The publication periods are as short as possible to keep pace with the speed of scientific advance, and in this respect the new books are comparable with scientific journals.

The first volume deals with dye lasers, the history of their development and their potential applications. Under the editorship of F. P. SCHÄFER six authors, all of whom have done pioneering work on dye lasers, review some major aspects. The editor introduces and discusses the principles of dye-laser operation in general, B. B. SNAVELY describes cw operation, and C. V. SHANK and E. P. IPPEN deal with mode-locking behavior. The basic laser material is a dye solution, and its chemical properties and structure are set forth by K. H. DREXHAGE. The broad spectrum of potential applications is outlined by T. W. Hänsch, who also touches briefly on the exciting field of laser spectroscopy, a most important application to which a forthcoming volume is devoted.

The treatment, being tutorial in nature, is suitable both for graduate students and for scientists working in the dye-laser field or applying a dye laser in another research discipline. The book will also prove to be an indispensable and handy source of information for the specialist. The literature is reviewed up to spring 1973, and the list of additional references (which cites the titles of articles) extends to summer 1973. This is proof of an amazingly short publication period for a 300-page book.

Heidelberg, October 1973   Helmut K. V. Lotsch

Contents

4. Structure and Properties of Laser Dyes. By K. H. Drexhage (with 6 Figures)

5. Applications of Dye Lasers. By Theodor W. Hänsch (with 26 Figures)

Contributors

DREXHAGE, K. H., Research Laboratories, Eastman Kodak Company, Rochester N. Y. 14650/USA

HÄNSCH, THEODOR W., Department of Physics, Stanford University, Stanford CA 94305/USA

IPPEN, E. P., Bell Telephone Laboratories, Holmdel N. J. 07733/USA

SCHÄFER, FRITZ P., Max-Planck-Institut für Biophysikalische Chemie, D-3400 Göttingen/Fed. Rep. of Germany

SHANK, C. V., Bell Telephone Laboratories, Holmdel N. J. 07733/USA

SNAVELY, BENJAMIN B., University of California, Lawrence Radiation Laboratory, Livermore, CA 24550/USA

1. Principles of Dye Laser Operation

Fritz P. Schäfer

With 53 Figures

Historical

Dye lasers entered the scene at a time when several hundreds of laser-active materials had already been found. Yet they were not just another addition to the already long list of lasers. They were the fulfillment of an experimenter's pipe dream that was as old as the laser itself: To have a laser that was easily tunable over a wide range of frequencies or wavelengths. Dye lasers are attractive in several other respects: Dyes can be used in the solid, liquid, or gas phases and their concentration, and hence their absorption and gain, is readily controlled. Liquid solutions of dyes are especially convenient: The active medium can be obtained in high optical quality and cooling is simply achieved by a flow system, as in gas lasers. Moreover, a liquid is self-repairing, in contrast to a solid-state active medium where damage (induced, say, by high laser intensities) is usually permanent. In principle, liquid dye lasers have output powers of the same magnitude as solid-state lasers, since the density of active species can be the same in both and the size of an organic laser is practically unlimited. Finally, the cost of the active medium, organic dyes, is negligibly small compared to that of solid-state lasers.

Early speculations about the use of organic compounds (RAUTIAN and SOBEL'MANN, 1961; BROCK et al., 1961) produced correct expectations of the role of vibronic levels of electronically excited molecules (BROUDE et al., 1963), but the first experimental study that might have led to the realization of an organic laser was by STOCKMAN et al. (1964) and STOCKMAN (1964). Using a high-power flashlamp to excite a solution of perylene in benzene between two resonator mirrors, STOCKMAN found an indication of a small net gain in his system. Unfortunately, he tried only the aromatic molecule perylene, which has high losses due to triplet-triplet absorption and absorption from the first excited singlet into higher excited singlet levels. Had he used some xanthene dye, like rhodamine 6 G or fluorescein, we would undoubtedly have had the dye laser two years earlier. In 1966, SOROKIN and LANKARD (1966) at IBM's Thomas J. Watson Research Center, Yorktown Heights, were the first to obtain stimulated emission from an organic compound, namely chloro-aluminum-phthalocyanine. Strictly speaking, this dye is an organometallic compound, for its central metal atom is directly bonded to an organic ring-type molecule, somewhat resembling to the compounds used in chelate lasers. In chelate lasers, however, stimulated emission originates in the central atom only, whereas in chloro-aluminum-phthalocyanine spectroscopic evidence clearly showed that the emission originated in the organic part of the molecule. SOROKIN and LANKARD set out to observe the resonance Raman effect in this dye, excited by a giant-pulse ruby laser (SOROKIN, 1969). Instead of sharp Raman lines they found a weak diffuse band at 755.5 nm, the peak of one of the fluorescence bands. They immediately suspected this might be a sign of incipient laser action and indeed, when the dye cell was incorporated into a resonator, a powerful laser beam at 755.5 nm emerged.

At that time the author, unaware of SOROKIN and LANKARD's work, was studying in his laboratory, then at the University of Marburg, the saturation characteristics of saturable dyes of the cyanine series. Instead of observing the saturation of the absorption, he used the saturation of spontaneous fluorescence excited by a giant-pulse ruby laser and registered by a photocell and a Tektronix 519 oscilloscope. The dye 3,3'-diethyltricarbocyanine had been chosen as a most convenient sample. This scheme worked well at very low concentrations of 10^{-6} and 10^{-5} mole/liter, but when VOLZE, then a student, tried to extend these measurements to higher concentrations, he obtained signals about one thousand times stronger than expected, with instrument-limited risetime that at a first glance were suggestive of a defective cable. Very soon, however, it became clear that this was laser action, with the four-percent reflection from the glass-air interface of the square, all-side-polished spectrophotometer cuvette acting as resonator mirrors. This was quickly checked by using a snooperscope to look at the bright spot where the infrared laser beam hit the laboratory wall. Together with Schmidt, then a graduate student, we photographed the spectra at various concentrations and also with reflective coatings added to the cuvette walls. Thus we obtained the first evidence that we had a truly tunable laser whose wavelength could be shifted over more than 60 nm by varying the concentration or the resonator mirror reflectivity. This was quickly confirmed and extended to a dozen different cyanine dyes. Among these was one that showed a relatively large solvatochromic effect and enabled us to shift the laser wavelength over 26 nm merely by changing the solvent (SCHÄFER et al., 1966). Stimulated emission was also reported in two other cyanine dyes by SPAETH and BORTFIELD at Hughes Aircraft Company (SPAETH and BORTFIELD, 1966). These authors, intrigued by SOROKIN and LANKARD's publication, used cryptocyanine and a similar dye excited by a giant-pulse ruby. Since their dyes had a very low quantum yield of fluorescence, they had a high-lying threshold and observed only a certain shift of laser wavelength with cell length and concentration.

Stimulated emission from phthalocyanine compounds, cryptocyanine, and methylene blue was also reported in 1967 by STEPANOV and co-workers (STEPANOV et al., 1967). They also reported laser emission from dyes that had a quantum efficiency of fluorescence of less than one thousandth of one percent; this, however, has not yet been confirmed by others.

A logical extension of this work was to utilize shorter pump wavelengths and other dyes in the hope of obtaining shorter laser wavelengths. This was first achieved in the author's laboratory by pumping a large number of dyes, among them several xanthene dyes, by the second harmonic of neodynium and ruby lasers (SCHÄFER et al., 1967). Similar results were obtained independently in several other laboratories (SOROKIN et al., 1967; MCFARLAND, 1967; STEPANOV et al., 1967; KOTZUBANOV et al., 1968). Dye laser wavelengths now cover the whole visible spectrum with extensions into the near-ultraviolet and infrared.

Another important advance was made when a diffraction grating was substituted for one of the resonator mirrors to introduce wavelength-dependent feedback (SOFFER and MCFARLAND, 1967). These authors obtained effective spectral narrowing from 6 to 0.06 nm and a continuous tuning range of 45 nm. Since then many different schemes have been developed for tuning the dye-laser wavelength; they will be discussed at length in the next chapter.

A natural step to follow was the development of flashlamps comparable in risetime and intensity to giant-pulse ruby lasers, to enable dye lasers to be pumped with a convenient, incoherent light source. This was initially achieved by techniques developed several years earlier for flash photolysis (SOROKIN and LANKARD, 1967; SCHMIDT and SCHÄFER, 1967). Soon after this, the author's team found that even normal linear xenon-filled flashlamps – and for some dyes even helical flashlamps – could be used, provided they were used in series with a spark gap and at sufficiently high voltage to result in a risetime of about one μsec. This is now standard practice for most single-shot or repetitively pumped dye lasers.

The common belief, that continuous-wave (cw) operation of dye lasers was not feasible because of the losses associated with the accumulation of dye molecules in the metastable triplet state, was corrected by SNAVELY and SCHÄFER (1969). Triplet-quenching by oxygen was found to decrease the steady-state population of the triplet state far enough to permit cw operation, at least in the dye rhodamine 6G in methanol solution, as demonstrated by using a lumped-parameter transmission line to feed a 500 μsec trapezoidal voltage pulse to a flashlamp, giving a 140-μsec dye-laser output. The premature termination of the dye laser pulse was also shown to be caused not by triplet accumulation but rather by thermal and acoustical schlieren effects. Our findings were later confirmed and extended by others who also used unsaturated hydrocarbons as triplet quenchers (MARLING et al., 1970a and b; PAPPALARDO et al., 1970a).

The pump-power density reqirements for a cw dye laser inferred from such measurements were so high that it seemed unlikely cw operation would be effected by pumping with the available high-power arc lamps. On the other hand, much higher pump-power densities can be obtained by focussing a high-power gas laser into a dye cuvette. The pumped region is thereby limited to about 50 µl, but the gain in the dye laser is usually very high.

Cw operation was a most important step towards the full utilization of the dye laser's potential; it was first achieved by PETERSON et al. (1970) at Eastman-Kodak Research Laboratory. They used an argon-ion laser to pump a solution of rhodamine 6 G in water with some detergent added. Water as solvent has the advantage of a high heat capacity, thus reducing temperature gradients which are further minimized by the high velocity of the flow of dye solution through the focal region. The detergent both acts as triplet quencher and prevents the formation of non-fluorescing dimers of dye molecules, which produce a high loss in pure water solutions.

This breakthrough triggered a host of investigations and developments of cw-dye lasers in the past two years; these will be covered in Chapter 2 of this book.

Towards the other end of the time scale, dye lasers, because of their extremely broad spectral bandwidth, hold great promise for the production of ultrashort pulses of smaller half-width than with any other laser. The first attempts to produce ultrashort pulses with dye lasers involved pumping a dye solution with a mode-locked pulse train from a solid-state laser; the dye cuvette sat in a resonator with a round-trip time exactly equal to, or a simple submultiple of, the spacing of the pump pulses (GLENN et al., 1968; BRADLEY and DURRANT, 1968; SOFFER and LINN, 1968). It subsequently proved possible to eliminate the resonator and use the superradiant traveling-wave emission from a wedged dye cuvette (MACK, 1969). The pulsewidths obtained in this way were generally some 10 to 30 psec. Self-mode-locking of flashlamp-pumped dye lasers was first achieved by SCHMIDT and SCHÄFER (1968), using rhodamine 6 G as the laser active medium and a cyanine dye as the saturable absorber. In an improved arrangement of this type a pulsewidth of only 2 psec was reached (BRADLEY et al., 1970). Eventually, the technique was applied to cw dye lasers to give a continuous mode-locked emission with pulses of only 1.5 psec (DIENES et al., 1972). This is still far from the theoretically possible limit. If mode-locking over the full bandwidth of about 250 THz were possible, the pulsewidth should be of the order of 5 fsec. Many more theoretical and experimental investigations will be necessary before we come anywhere near this limit. Chapter 3 of this book is devoted to this important field of dye laser research.

If one tries to extrapolate this historical survey and discern the outline of future developments, one can immediately foresee large quantitative improvements in many features: The wavelength coverage will be extended farther into the ultraviolet, peak pulse powers and energies will increase by several orders of magnitude, the output power and frequency stability of cw dye lasers will also be improved by orders of magnitude, and the generation of ultrashort pulses of unprecedentedly small pulsewidths will become possible. A qualitative improvement will be the incoherent pumping of cw dye lasers with specially developed high-power arc lamps that will allow a much higher output power than pumping with gas lasers. A most important development will concern the chemical aspect of the dye laser, namely the synthesis of special dyes, a field of research that has just been opened (DREXHAGE, 1972a).

Many developments will undoubtedly be sparked off by the demands of the many and various applications of lasers. Their number is constantly growing and even today they are already so numerous that Chapter 5 of this book, devoted to applications, can hardly encompass them all.

Organization of the Book

Chapter 1 by SCHÄFER is introductory. The chemical and spectroscopic properties of organic compounds are described, there is a tutorial presentation of the general principles of dye-laser operation, and aspects of the dye laser not treated in one of the special chapters are reviewed. Emphasis here is on an easy understanding of the physical and chemical principles involved, rather than on completeness, or historical or systematic presentation. On this basis the specialized chapters will be built up.

Chapter 2 by SNAVELY presents a review of cw dye lasers, with emphasis on the gain analysis. Special attention is given to the triplet problem.

Mode-locking of dye lasers is the topic of Chapter 3 by SHANK and IPPEN. After an introduction covering methods of measuring ultrashort pulses, the authors discuss experimental methods applicable to mode-locking, pulsed and cw dye lasers. The authors make it clear that we still do not understand the mechanism of mode-locking in dye lasers.

The chemical and physical properties of dyes are discussed in Chapter 4 by DREXHAGE. A large amount of data on dyes is presented in support of the rules found by the author for the selection or synthesis of useful laser dyes. A list of laser dyes completes this chapter.

Finally, Chapter 5 on applications of dye lasers by Hänsch gives an idea of the broad range of today's applications and offers perspectives for future developments.

The authors have tried to make the chapters as self-contained as possible, so that they can be read independently, albeit at the cost of some slight overlap.

1.1. General Properties of Organic Compounds

Organic compounds are defined as hydrocarbons and their derivatives. They can be subdivided into saturated and unsaturated compounds. The latter are characterized by the fact that they contain at least one double or triple bond. These multiple bonds not only have a profound effect on chemical reactivity, they also influence spectroscopic properties. Organic compounds without double or triple bonds usually absorb at wavelengths below 160 nm, corresponding to a photon energy of 180 kcal/mole. This energy is higher than the dissociation energy of most chemical bonds, therefore photochemical decomposition is likely to occur, so such compounds are not very suitable as the active medium in lasers. In unsaturated compounds all bonds are formed by σ electrons; these are characterized by the rotational symmetry of their wave function with respect to the bond direction, i.e. the line connecting the two nuclei that are linked by the bond. Double (and triple) bonds also contain a σ bond, but in addition use π electrons for bonding. The π electrons are characterized by a wave function having a node at the nucleus and rotational symmetry along a line through the nucleus and normal to the plane subtended by the orbitals of the three σ electrons of the carbon or heteroatom (Fig. 1.1). A π bond is formed by the lateral overlap of the π-electron orbitals, which is maximal when the symmetry axes of the orbitals are parallel. Hence, in this position, bond energy is highest and the energy of the molecule minimal, thus giving a planar molecular skeleton of high rigidity. If two double bonds are separated by a single

bond, as in the molecule butadiene,
$$\begin{array}{ccc} H & & H \\ \diagdown & & \diagup \\ C{=}C{-}C{=}C \\ \diagup & & \diagdown \\ H & & H \end{array}$$
, the two

double bonds are called *conjugated*. Compounds with conjugated double bonds also absorb light at wavelengths above 200 nm. All dyes in the proper sense of the word, meaning compounds having a high absorption in the visible part of the spectrum, possess several conjugated double

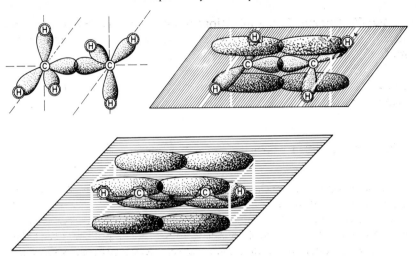

Fig. 1.1. Orbitals and bonds in ethane, ethylene, and acetylene

bonds. The basic mechanism responsible for light absorption by compounds containing conjugated double bonds is the same, in whatever part of the spectrum these compounds have their longest wavelength absorption band, whether near-infrared, visible, or near-ultraviolet. We thus use the term *dye* in the wider sense as *encompassing all substances containing conjugated double bonds*. Whenever the term *dye* is used in this book, it will have this meaning.

For the remainder of this chapter we restrict our discussion of the general properties of organic compounds to dyes, since for the foreseeable future these are the only organic compounds likely to be useful laser-active media.

The thermal and photochemical stability of dyes is of utmost importance for laser applications. These properties, however, vary so widely with the almost infinite variety of chemical structure, that practically no generally valid rules can be formulated. Thermal stability is closely related to the long-wavelength limit of absorption. A dye absorbing in the near-infrared has a low-lying excited singlet state and, even slightly lower than that, a metastable triplet state. The triplet state has two unpaired electrons and thus, chemically speaking, biradical character. There is good reason to assume that most of the dye molecules that reach this highly reactive state by thermal excitation will react with solvent molecules, dissolved oxygen, impurities, or other dye molecules to yield decomposition products. The decomposition would be of pseudo-first order with a reaction constant $k_1 = A \exp(-E_A/RT)$, where A is the Arrhenius constant and has most often a value of 10^{12} sec^{-1} for reactions of this type (ranging from 10^{10} to 10^{14} sec^{-1}), E_A is the activation energy, R is the gas constant and T the absolute temperature. The half-life of such a dye in solution then is $t_{1/2} = \ln 2/k_1$. Assuming as a minimum practical lifetime one day, the above relations yield an activation energy of 24 kcal/mole, corresponding to a wavelength of 1.2 μm. If $A = 10^{10}$ sec^{-1}, this shifts the wavelength to 1.7 μm, and with $A = 10^{14}$ sec^{-1} it would correspond to 1.1 μm. If we assume that a year is the minimum useful half-life of the dye (and $A = 10^{12}$ sec^{-1}), we get a wavelength of 1.0 μm.

Obviously, it becomes more and more difficult to find stable dyes having the maximum of their long-wavelength band of absorption in the infrared beyond 1.0 μm, and there is little hope of ever preparing a dye absorbing beyond 1.7 μm that will be stable in solution at room temperature. Thus, dye-laser operation at room temperature in the infrared will be restricted to wavelengths not extending far beyond 1.0 μm.

The short-wavelength limit of dye-laser operation, already mentioned implicitly, is given by the absorption of dyes containing only two conjugated double bonds and having their long-wavelength absorption band at wavelengths of about 220 nm. Since the fluorescence, and hence the laser emission, is always red-shifted, dye lasers can hardly be expected to operate at wavelengths below about 250 nm. Even if we were to try to use compounds possessing only one double bond, like ethylene, absorbing at 170 nm, we could at best hope to reach 200 nm in laser emission. At this wavelength, however, photochemical decomposition already competes effectively with radiative deactivation of the molecule, since the energy of the absorbed quantum is higher than the energy of any bond in the molecule.

Another important subdivision of dyes is into ionic and uncharged compounds. This feature mainly determines melting point, vapor pressure, and solubility in various solvents. An uncharged dye already mentioned is butadiene, $CH_2{=}CH{-}CH{=}CH_2$; other examples are most aromatics: anthracene, pyrene, perylene, etc. They usually have relatively low-lying melting points, relatively high vapor pressures, and good solubility in nonpolar solvents, like benzene, octane, cyclohexane, chloroform, etc. Cationic dyes include the large class of cyanine dyes, e.g.

the simple cyanine dye
$$\left[\begin{array}{c} CH_3 \quad\quad H \quad\quad CH_3 \\ \overset{+}{N} \quad C \quad N \\ CH_3 \quad C \quad C \quad CH_3 \\ H \quad H \end{array} \right]^{+} J^{-}.$$
These

compounds are salts, consisting of cations and anions, so they have high melting points, very low vapor pressure, good solubility in more polar solvents like alcohols, and only a slight solubility in less polar solvents. Similar statements can be made for anionic dyes, e.g. for the dye

Many dyes can exist as cationic, neutral and anionic molecules depending on the pH of the solution, e.g. fluorescein:

Neutral form	Cationic form	Di-anionic form
(alcoholic solution)	(hydrochloric acid solution)	sodium hydroxyde solution

It should be stressed here that dyes are potentially useful as laser-active media in the solid, liquid and vapor phases. The last has not yet been used in a dye laser, but preliminary experiments in several laboratories look promising. Since most dyes form good single crystals, it might be attractive to use them directly in this form. There are two main obstacles to their use in crystal form: The extremely high values attained by the extinction coefficient in dyes, which prevents the pump-light from exciting more than the surface layer a few microns thick; and the concentration quenching of fluorescence that usually sets in whenever the dye molecules approach each other closer than about 10 nm. Doping a suitable host crystal with a small fraction (one thousandth or less) of dye circumvents these difficulties. On the other hand, solid solutions of many different types can be used. For instance, one can dissolve a dye in the liquid monomer of a plastics material and then polymerize it; or one can dissolve it in an inorganic glass (e.g. boric acid glass) or an organic glass (e.g. sucrose glass) or some semirigid material like gelatine or polyvinylalcohol. The utilization of some of these techniques for dye lasers will be discussed later, together with the use of dyes in the liquid and vapor phases.

1.2. Light Absorption by Organic Dyes

The light absorption of dyes can be understood on a semiquantitative basis if we take a highly simplified quantum-mechanical model, such as the free-electron gas model (KUHN, 1958–59). This model is based on the fact that dye molecules are essentially planar, with all atoms of the conjugated chain lying in a common plane and linked by σ bonds. By comparison, the π electrons have a node in the plane of the molecule and form a charge cloud above and below this plane along the conjugated chain. The centers of the upper and lower lobes of the π-electron cloud are about one half bond length distant from the molecular plane. Hence, the electrostatic potential for any single π electron moving in the field of the rest of the molecule may be considered constant, provided all bond lengths and atoms are the same (Fig. 1.2). Assume, that the conju-

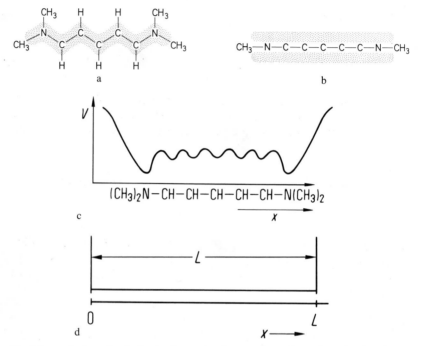

Fig. 1.2. a π-electron cloud of a simple cyanine dye seen from above the molecular plane; b the same as seen from the side; c potential energy V of a π-electron moving along the zig-zag chain of carbon atoms in the field of the rump molecule; d simplified potential energy trough; L = length of the π-electron cloud in a as measured along the zig-zag chain. (From FÖRSTERLING and KUHN, 1971)

gated chain which extends approximately one bond length to the left and right beyond the terminal atoms has length L. Then the energy E_n of the nth eigenstate of this electron is given by $E_n = h^2 n^2 / 8mL^2$, where h is PLANCK's constant, m is the mass of the electron, and n is the quantum number giving the number of antinodes of the eigenfunction along the chain. According to the Pauli principle, each state can be occupied by two electrons. Thus, if we have N electrons, the lower $1/2N$ states are filled with two electrons each, while all higher states are empty (provided N is an even number; this is usually the case in stable molecules since only highly reactive radicals possess an unpaired electron). The absorption of one photon of energy $\Delta E = hc_0/\lambda$ (where λ is the wavelength of the absorbed radiation and c_0 is the velocity of light) raises one electron from an occupied to an empty state. The longest wavelength absorption band then corresponds to a transition from the highest occupied to the lowest empty state with

$$\Delta E_{min} = \frac{h^2}{8mL^2}(N+1) \quad \text{or} \quad \lambda_{max} = \frac{8mc_0}{h} \frac{L^2}{N+1}.$$

This indicates that to first approximation the position of the absorption band is determined only by the chain length and by the number of π electrons N. Good examples of this relation are the symmetrical cyanine dyes of the general formula

where j is the number of conjugated double bonds, R_1 a simple alkyl group like C_2H_5, and R indicates that the terminal nitrogen atoms are part of a larger group, as e.g. in the following dye (homologous series of thiacyanines):

The double-headed arrow means that the two formulae are limiting structures of a resonance hybrid. The π electrons in the phenyl ring can be neglected in first approximation, or treated as a polarizable charge cloud, leading to an apparent enlargement of the chain L. In the case of the last-mentioned dye, good agreement is found between calculated and experimental absorption wavelength, when the chain length L is assumed to extend 1.3 bond lengths (instead of 1.0 bond length as above) beyond the terminal atoms. The bond length in cyanines is 1.40 Å. The good agreement between the results of this simple calculation and the experimental data for the above thiacyanines is shown by the following comparison:

Wavelength (in nm) of absorption maximum for thiacyanines

	Number of conjugated double bonds $j =$			
	2	3	4	5
Calculated	395	521	649	776
Experimental	422	556	652	760

Similarly good agreement can be found for all the other homologous series of symmetrical cyanines, once the value of the end length extending over the terminal N atoms is found by comparison with the experimentally observed absorption wavelength for one member of the series. Absorption wavelengths have been reported recently for a large number of cyanines (Miyazoe and Maeda, 1970).

The following nomenclature is customarily used for cyanine dyes. If $j = 2$, then the dye is a cyanine in the narrower sense or monomethine dye, since it contains one methine group, —CH—, in its chain; if $j = 3$, the dye is a carbocyanine or trimethine dye; if $j = 4, \ldots, 7$, the dye is a di-, tri-, tetra-, or pentacarbocyanine or penta-, hepta-, nona- or undecamethine dye. The heterocyclic terminal groups are indicated in an abridged notation; thus "thia" stands for the benzthiazole group:

and "oxa" for the benzoxazole group

In these groups the atoms are numbered clockwise, starting from the sulfur or oxygen atom. Thus, the name of the last-mentioned thiacyanine dye is 3,3'-diethyl-thiadicarbocyanine for $j = 4$. If the terminal group is quinoline, the syllable "quinolyl" indicative of this end group is frequently omitted from the name. In this case, however, the position of linkage of the polymethine chain to the quinoline ring must be given, e.g.

is 1,1'-diethyl-2,2'-carbocyanine (trivial name: pinacyanol), while

is 1,1'-diethyl-2,4'-carbocyanine (trivial name: dicyanine), and

is 1,1'-diethyl-4,4'-carbocyanine (trivial name: cryptocyanine). Where there are two different end groups, these are named in alphabetical order, e.g. 3,3'-diethyl-oxa-thiacarbocyanine is

It is easy to eliminate the above assumption of identical atoms in the conjugated chain. If, for instance, one CH group is replaced by an N atom, the higher electronegativity of the nitrogen atom adds a small potential well to the constant potential; a simple perturbation treatment shows that every energy level is shifted by $\varepsilon = -B\Psi^2$, where B is a constant characteristic of the electronegativity of the heteroatom (in the case of $=N-$, $B = 3.9 \times 10^{-20}$ erg · cm) and Ψ is the value of the normalized wave function at the heteroatom. This means that the shift of an energy level is zero if the wave function has a node at the heteroatom, and maximal if it has an antinode there. The change in absorption wavelength with heterosubstitution calculated in this way is in good agreement with the experimentally observed values. The second of the above assumptions, that of equal bond lengths along the chain, can be eliminated in a similar way. An illustrative example is offered by the polyenes of the general formula $R-(CH=CH)_j-R$. These compounds have an even number of atoms. The total charge distribution results in localized double and single bonds (Fig. 1.3) and hence in alternating short and long bond lengths (1.35 Å and 1.47 Å, respectively). A π electron moving along the chain will therefore experience greater attraction to the neighboring atoms in the middle of a double bond than in the middle of a single bond. This fact may be represented by a periodic perturbing potential with minima at the center of the double bonds and maxima at the center of the

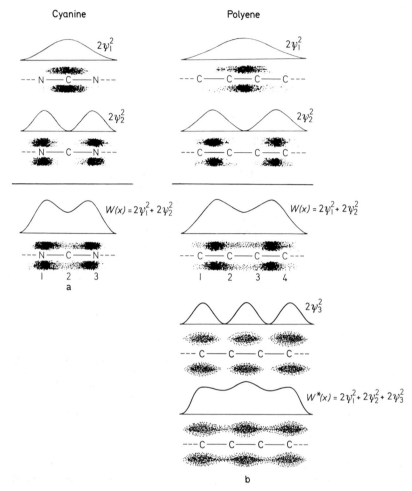

Fig. 1.3. Electron densities (proportional to the square of the wave function, ψ^2) along the conjugated chain of a) a cyanine, b) a polyene (butadiene). The symbols ψ_1 and ψ_2 denote the wave functions of the first and second filled π-electron molecular orbital, and ψ_3 that of the lowest orbital normally empty, which is occupied by one electron when excited. $W(x)$ is the total π-electron density for a molecule in the ground state, and $W^*(x)$ is the same for butadiene in the first excited state

single bonds (Fig. 1.4). Assuming a sinusoidal potential with an amplitude of 2.4 eV gives good agreement between the calculated and experimental values of the absorption wavelengths. The relation between chain length and absorption wavelength is found to be very different in the cyanines and the polyenes. In symmetrical cyanines the absorption wavelength is

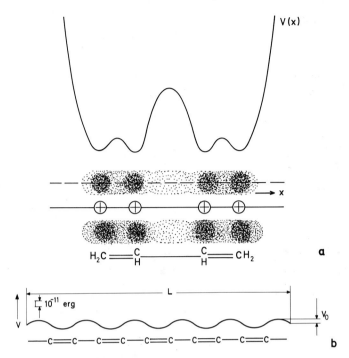

Fig. 1.4. a Potential energy $V(x)$ of a π-electron moving along the carbon atom chain in the field of the rump molecule of butadiene; b simplified potential energy trough of a long polyene molecule with perturbing sinusoidal potential of amplitude $V_0 = 2.4$ eV. The energy difference between highest filled and lowest empty orbital is given by $\Delta E = (h^2/8mL^2)(2j+1) + 0.83(1-1/2j)V_0$, where j is the number of conjugated double bonds. (From KUHN, 1959)

shifted by a constant amount of roughly 100 nm on going from one member of a series to the next higher which has one more double bond. In polyenes this shift decreases with increasing number of double bonds. A similar treatment can be applied to unsymmetrical cyanines in which the difference in electronegativity of the different end groups gives a similar polyene-type perturbation. The perturbation is less pronounced when the difference in electronegativity of the end groups is less.

Many dyes containing a branched chain of conjugated double bonds can, in a first approximation, be classified either as symmetrical cyanine-like or polyene-like substances, or as intermediate cases. For example, Michler's hydrol blue absorbs at practically the same wavelength as the symmetrical cyanine

$$(H_3C)_2\bar{N}\!-\!(CH\!=\!CH)_4\!CH\!=\!\overset{\oplus}{N}(CH_3)_2$$

$$\leftrightarrow (H_3C)_2\overset{\oplus}{N}\!=\!CH\!-\!(CH\!=\!CH)_4\bar{N}(CH_3)_2.$$

Hence it seems justified to treat it like a cyanine by neglecting the weakly printed bonds in the formula of Michler's hydrol blue. If, however, additional branching is introduced by connecting positions 4 and 8 through a $\left(\begin{array}{c} =\!N\!-\! \\ | \\ CH_3 \end{array}\right)$ bridge, this gives acridine orange

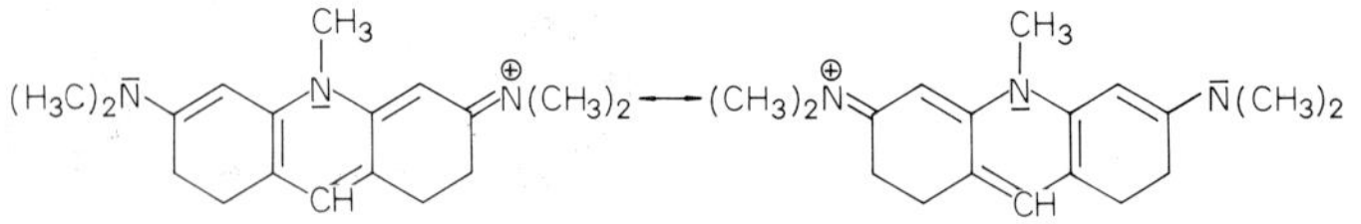

The absorption wavelength is shifted from 603 nm to 491 nm. It indicates that in this case branching cannot be neglected even in a first approximation.

With an O atom as bridging group instead of the N—CH$_3$ group, the xanthylium dye pyronine G with an absorption wavelength of 550 nm is obtained:

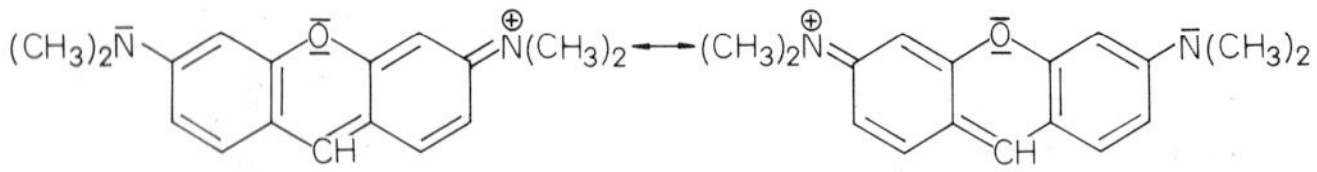

For a detailed treatment of molecules containing such a branched free-electron gas, the reader is referred to the literature (Kuhn, 1958/59).

Another important class of dyes in which the branching of the conjugated chain can be neglected to first approximation, are the phthalocyanines and similar large-ring molecules. In first approximation the benzene rings are separated resonance systems, leaving the 16-membered ring indicated in Fig. 1.5 by heavy lines. Now there are 18 π electrons on a ring of circumference $L = 16\,l$. In the lowest state ($n = 0$) the eigenfunction has no nodes, for $n \neq 0$ there are two degenerate eigenfunctions for every n, corresponding to the sine and cosine in the constant-potential approximation. A large perturbation is, however, introduced by the nitrogen atoms, which remove the degeneracy. This effect is most pronounced for $n = 4$, since one eigenfunction then has its nodes at the N atoms, the other its antinodes. The lowest unfilled levels are again degenerate because of equal perturbation energy. The absorp-

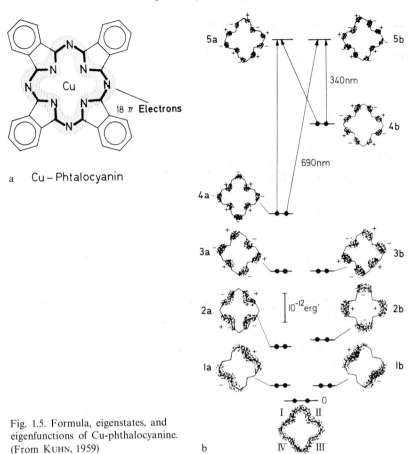

a   Cu – Phtalocyanin

Fig. 1.5. Formula, eigenstates, and
eigenfunctions of Cu-phthalocyanine.
(From Kuhn, 1959)

tion wavelengths for the two transitions from $n=4$ to $n=5$ are then readily calculated (with the above value for B): $\lambda_1 = 690$ nm and $\lambda_2 = 340$ nm (Fig. 1.5). The experimental values are 674 nm and 345 nm.

The free-electron model, even in its simplest form, gives very satisfactory agreement between calculated and experimental values of the absorption wavelength for large dye molecules. Nevertheless, its one-electron functions are not sufficient for a quantitative description of light absorption by small molecules of high symmetry, like benzene, naphthalene and similar molecules, since here the repulsion between the π electrons plays an important role. For the inclusion of electron correlations into the free electron model, the reader is referred to Försterling et al. (1966).

The oscillator strength of the absorption bands can also be calculated easily by the free-electron model. Transition moments X and Y along and normal to the long molecular axis are connected with the oscillator strength f of the absorption band by:

$$f = 2 \cdot \frac{8m_0\pi^2}{3h^2} \cdot \Delta E(X^2 + Y^2),$$

and can be calculated using the electron gas wave functions ψ_a and ψ_b of the eigenstates between which the transition occurs:

$$X = \int_{\text{molecule}} \psi_a x \psi_b dx, \qquad Y = \int_{\text{molecule}} \psi_a y \psi_b dy.$$

The relative strengths of the x and y component also yield the orientation of the transition moment in the molecule. One finds good agreement on comparing the f values obtained from the absorption spectrum of the dye using the relation

$$f = 4.32 \cdot 10^{-9} \int_{\substack{\text{absorption} \\ \text{band}}} \varepsilon(\tilde{v}) d\tilde{v},$$

where ε is the numerical value of the molar decadic extinction coefficient measured in liter/(cm $\cdot$ mole), and $\tilde{v}$ is the numerical value of the wave number measured in cm^{-1}.

A peculiarity of the spectra of organic dyes as opposed to atomic and ionic spectra is the width of the absorption bands, which usually covers several tens of nanometers. This is immediately comprehensible when one recalls that a typical dye molecule may possess fifty or more atoms, giving rise to about 150 normal vibrations of the molecular skeleton. These vibrations, together with their overtones, densely cover the spectrum between a few wave numbers and $3000 \, cm^{-1}$. Many of these vibrations are closely coupled to the electronic transitions by the change in electron densities over the bonds constituting the conjugated chain. After the electronic excitation has occurred, there is a change in bond length due to the change in electron density. If, e.g. in Fig. 1.3, the bond length connecting atoms 1 and 2 is r (typically 1.35 Å) and this bond is lengthened in the excited molecule, the new equilibrium distance being r^*, atoms 1 and 2 will start to oscillate, classically speaking, around this new position with an amplitude $r^* - r$ (typically 0.02 Å) after the electronic transition has occurred. A molecular skeletal vibration is excited in this way. The new equilibrium total π electron density $W^*(x)$ for the excited state is given in Fig. 1.3c. demonstrating the large increase in π electron density over the bond connecting atoms 2 and 3. Quantum-mechanically this means that transitions have occurred from the electronic and vibrational ground state S_0 of the molecule to an electronically and vibrationally excited state S_1, as depicted in Fig. 1.6. This results in spectra like Fig. 1.6b or c, depending on how many of the vibrational sublevels, spaced at $h v_0 \cdot (v + 1/2)$, with $v = 0, 1, 2, 3 \dots$, are reached and what the transition moments to these sublevels are.

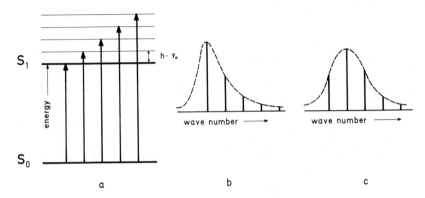

Fig. 1.6. a Electronic and vibronic energy levels of a dye molecule; S_0: ground state, S_1: excited state, and absorptive transitions from S_0 to S_1; b and c two possible forms of concomitant spectra

In the general case of a large dye molecule, many normal vibrations of differing frequencies are coupled to the electronic transition. Furthermore, collisional and electrostatic perturbations, caused by the surrounding solvent molecules broaden the individual lines of such vibrational series as that given in Fig. 1.6. As a further complication, every vibronic sublevel of every electronic state, including the ground state, has superimposed on it a ladder of rotationally excited sublevels. These are extremely broadened because of the frequent collisions with solvent molecules which hinder the rotational movement so that there is a quasicontinuum of states superimposed on every electronic level. The population of these levels in contact with thermalized solvent molecules is determined by a Boltzmann distribution. After an electronic transition, which, as described above, leads to a nonequilibrium state (Franck-Condon state) the approach to thermal equilibrium is very fast in liquid solutions at room temperature. The reason is that a large molecule experiences at least 10^{12} collisions/sec with solvent molecules, so that equilibrium is reached in a time of the order of one picosecond. Thus the absorption is practically continuous all over the absorption band. The same is true for the fluorescence emission corresponding to the transition from the electronically excited state of the molecule to the ground state.

This results in a mirror image of the absorption band displaced towards lower wave numbers by reflection at the wave number of the purely electronic transition. This condition exists, since the emissive transitions start from the vibrational ground state of the first excited electronic state S_1 and end in vibrationally excited sublevels of the electronic ground state. The resulting typical form of the absorption and fluorescence spectrum of an organic dye is given in Fig. 1.7.

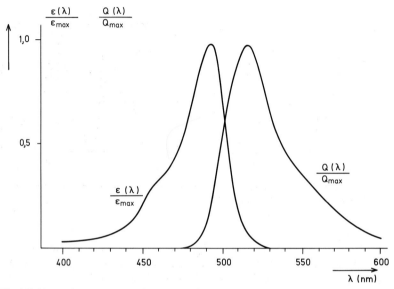

Fig. 1.7. Absorption spectrum, $\varepsilon(\lambda)/\varepsilon_{max}$, and fluorescence spectrum, $Q(\lambda)/Q_{max}$, of a typical dye molecule (fluorescein-Na in water)

Further complications of dye spectra arise from temperature and concentration dependence and acid–base equilibria with the solvent. If the temperature of a dye solution is increased, higher vibrational levels of the ground state are populated according to a Boltzmann-distribution, and more and more transitions occur from these to higher sublevels of the first excited singlet state. Consequently, the absorption spectrum becomes broader and the superposition of so many levels blurs most of the vibrational fine structure of the band, while cooling of the solution usually reduces the spectral width and enhances any vibrational features that may be present. Thus, spectra of solid solutions of dyes in EPA, a mixture of 5 parts ethyl ether, 5 parts isopentane, and 5 parts ethanol that forms a clear organic glass when cooled down to 77 °K, are often used for comparison with calculated spectra because of their well-resolved vibrational structure. Further cooling below the glass point, when the free movement of solvent molecules or parts thereof is inhibited, usually brings about no further sharpening of the spectral features (Fig. 1.8). The many possible different configurations of the solvated molecule in the cage of solvent molecules cannot be attained when the temperature is lowered because the activation energy required to reach the new equilibrium positions is too high. A very special case is the

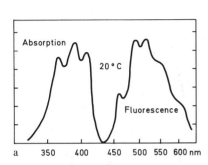
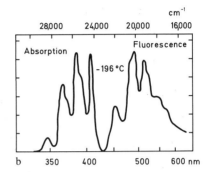

Fig. 1.8. Absorption and fluorescence spectra of diphenyloctatetraene in xylene. a at 20° C, b at −196° C. (From HAUSSER et al., 1935)

Shpolski effect (SHPOLSKI, 1962): This refers to the appearance of very sharp, line-like spectra (often termed quasi-line spectra in the Russian literature) of about one cm^{-1} width instead of the usual diffuse band spectra of dye molecules (most often aromatics) in a matrix of n-paraffins at low temperature (usually below 20 °K). Evidently this is because there are only few different possibilities of solvation of the molecule in that matrix, and each of the different sites causes a series of spectral lines in absorption as well as in emission. Analyzing these series gives the energy difference of the different sites, usually a few hundred cm^{-1} (Fig. 1.9). The Shpolski effect has also been called the optical analog of the Möss-bauer effect (REBANE and KHIZHNYAKOV, 1963), since here also the recoil of the interacting photons seems to be taken up by the matrix as a whole.

The concentration dependence of dye spectra is most pronounced in solutions where the solvent consists of small, highly polar molecules, notably water. Dispersion forces between the large dye molecules tend to bring the dye molecules together in a position with the planes of the molecules parallel, where the interaction energy usually is highest. This is counteracted by the repulsive Coulomb forces if the dye molecules are charged. In solvents of high dielectric constants this repulsion is lowered and the monomer–dimer equilibrium is far to the side of the dimer. The

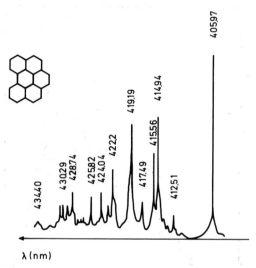

λ (nm)

Fig. 1.9. Fluorescence spectrum of 1,12-benzoperylene in n-hexane at 77° K (Shpolski-effect). (From PERSONOV and SOLODUNOV, 1967)

equilibrium constant of the dimerization process monomer + monomer $\rightleftharpoons$ dimer is $K = [\text{dimer}]/[\text{monomer}]^2$ and can easily be obtained from spectra of differently concentrated dye solutions (Fig. 1.10). Usually the polymerization process stops at this stage, as evidenced by at least one isosbestic point (535 nm in Fig. 1.10). However, in some cases reversible polymerization occurs to a degree of polymerization of up to 1 million dye molecules (SCHEIBE, 1941; BÜCHER and KUHN, 1970). In this case, a very sharp, high-intensity absorption and emission line becomes prominent in the spectra. The spectral differences between monomer and dimer, or even polymer, dye molecules can easily be understood at least qualitatively (FÖRSTER, 1951). Figure 1.11 shows schematically the two main electronic energy levels of two distant dye molecules, and the splitting of the degenerate levels if they come close enough to each other to experience interaction energy and the concomitant formation of a dimer. Now two transitions are possible, and these in general possess different transition moments depending on the wave functions of the dimer. Most often the long-wavelength transition has a practically vanishing transition moment, so that only one absorption band of the dimer is observed, lying to the short-wavelength side of the monomer band. This has an important consequence for the fluorescence of such dimers. Since the upper level from which the fluorescence starts is always the lowest-lying excited electronic level, and since a small transition moment is coupled to a long lifetime of the excited state, these dimers

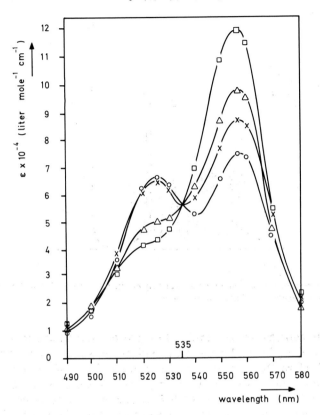

Fig. 1.10. Absorption spectra of aqueous solutions of rhodamine B at 22° C, ○ 1.5 × 10⁻³ M, × 7.6 × 10⁻³ M, Δ 1.5 × 10⁻⁴ M, □ 3.0 × 10⁻⁶ M. (From SELWYN and STEINFELD, 1972)

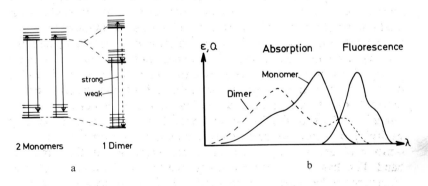

Fig. 1.11. a Energy levels of two monomers and the dimer molecule formed by them; b resulting spectra

would show a very slow decay of their fluorescence. This, however, makes them susceptible to competing quenching processes, which in liquid solution are generally diffusion-controlled and hence very fast processes. Consequently, in most of these cases the fluorescence of the dimers is completely quenched and cannot be observed.

This is the reason why dimers constitute an absorptive loss of pump power in dye lasers and must be avoided by all means. There are several ways in which this may be done. One is to use a less polar solvent, like alcohol or chloroform. There are very few dyes which show dimerization in alcohol at the highest concentrations and at low temperatures. Another possibility is to add a detergent to the aqueous dye solution, which then forms micelles that contain one dye molecule each (FÖRSTER and SELINGER, 1964). This method is of prime importance for cw dye lasers and is discussed in Chapter 4, on laser dyes.

Acid–base equilibria have already been mentioned above for the case of fluorescein. Very often these equilibria are less obvious. For example, the dye 3,6-bis-dimethylaminoacridine can be protonated at the central nitrogen atom and then exhibits an absorption band shifted to longer wavelengths by about 55 nm (FERGUSON and MAU, 1972b), as shown in Fig. 1.12. Several other cases of acid–base equilibria of laser dyes will be discussed in detail in Chapter 4.

In addition to above-mentioned transitions between σ orbitals ($\sigma \to \sigma^*$ transitions) and the extensively discussed transitions between

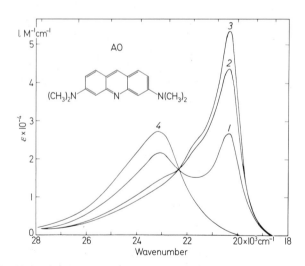

Fig. 1.12. Absorption spectra of 3.6-bis-dimethyl-aminoacridine (AO) in ethanol, 1 3×10^{-5} M, 2 2% water added, 3 bubbled with CO_2 gas, 4 anhydrous potassium carbonate added. (From FERGUSON and MAU, 1972b)

π orbitals ($\pi \to \pi^*$ transitions), we should also mention the transitions from the orbital of a lone-electron pair (so-called n orbitals), e.g. in a keto group, $C = O$, to a π orbital. Since the n orbitals have either spherical or rotational symmetry, with the symmetry axis lying in the molecular plane, the overlap with π orbitals, and hence the transition moments, are very small. Correspondingly, the molar decadic extinction coefficient ε (as defined by BEER's law, $I = I_0 \cdot 10^{-\varepsilon \cdot c \cdot d}$, where I and I_0 are the transmitted and incident light intensity, c is the concentration in mole/liter of the absorbing species, and d is the thickness of the absorbing layer in cm) of absorption bands caused by n $\to \pi^*$ transitions usually ranges from 1 to 10^3 l/(mole $\cdot$ cm), while $\pi \to \pi^*$ transitions exhibit values of ε lying between 10^3 and 10^6 l/(mole $\cdot$ cm).

The free-electron model can also provide a simple explanation for another important property of the energy levels of organic dyes, namely the position of the triplet levels relative to the singlet levels. In the ground state of the dye molecule, which in the free-electron model is the state with the $(1/2)N$ lowest levels filled, the spins of two electrons occupying the same level are necessarily antiparallel, resulting in zero spin. However, there is also the possibility of a parallel arrangement of the spins if one of the electrons is raised to a higher level. The resulting spin $S = 1$ can place itself either parallel, antiparallel or orthogonal with respect to an external magnetic field. The parallel arrangement of the spins of the two most energetic electrons thus gives a triplet state of the same energy as the singlet state with zero spin within the framework of one-electron functions. The Dirac formulation of the Pauli exclusion principle states that the total wave function, including the spin function, must be antisymmetric with respect to the exchange of any two electrons. In the two-electron case considered this means that the following four-antisymmetrical product wave function can be used. Here spin $+ 1/2$ is described by the spin function α, spin $- 1/2$ by the spin function β:

$$\psi_s = \{\psi_m(1)\psi_n(2) + \psi_n(1)\psi_m(2)\} \{\alpha(1)\beta(2) - \alpha(2)\beta(1)\} \, ,$$

$$\psi_{T, +1} = \{\psi_m(1) \Psi_n(2) - \psi_n(1)\psi_m(2)\} \{\alpha(1)\alpha(2)\} \, ,$$

$$\psi_{T, 0} = \{\psi_m(1)\psi_n(2) - \psi_n(1)\psi_m(2)\} \{\alpha(1)\beta(2) + \alpha(2)\beta(1)\} \, ,$$

$$\psi_{T, -1} = \{\psi_m(1)\psi_n(2) - \psi_n(1)\psi_m(2)\} \{\beta(1)\beta(2)\} \, ,$$

where ψ_s is the singlet, $\psi_{T,+1}, \psi_{T,-1}, \psi_{T,0}$ are the three triplet wave functions, and the argument 1 or 2 refers to electron no. 1 or no. 2. Because of the symmetry of the spin factor, the spatial factor of these functions is symmetric for the singlet wavefunction and antisymmetric for the triplet wave functions. These spatial factors of one-dimensional two-electron functions can be interpreted in terms of two-dimensional one-electron functions

$$\psi_{m,n}(s_1,s_2) + \psi_{n,m}(s_1,s_2)$$

for the singlet case and

$$\psi_{m,n}(s_1,s_2) - \psi_{n,m}(s_1,s_2)$$

for the triplet case.

In a crude approximation, one can think of electron *1* as travelling in the upper lobe of the electron cloud along the molecular chain its coordinate being s_1, and electron *2* in the lower lobe its coordinate being s_2, as depicted for three relative positions in the right half of Fig. 1.13a. In the left half of this figure these three positions A, B, C are plotted in a plane determined by s_1 and s_2 as cartesian coordinates.

For every configuration of the two electrons we can give the repulsion energy of the two electrons

$$V = \frac{e_0^2}{Dr} = \frac{e_0^2}{D[(s_1 - s_2)^2 + d^2]^{1/2}},$$

where r is the distance between the two electrons, $d = 0.12$ nm is the distance between the centers of the upper and lower lobes, and D denotes the dielectric constant in which the two electrons are imbedded.

The potential energy profile has a crest along the symmetry axis $s_1 = s_2$(Fig. 1.13b). Since the spatial factor of the singlet function is symmetric with respect to this axis, it must have antinodes there. By contrast, the antisymmetric spatial factor of the triplet wave functions has a nodal line along $s_1 = s_2$. Consequently, the mean potential energy of the electrons in the singlet state with quantum numbers n,m is higher than those in the triplet state with the same quantum numbers. This crude model gives the most important result that *for every excited singlet state there exists a triplet state of somewhat lower energy.* In addition the numerical calculation usually yields a value of the energy difference between corresponding singlet and triplet states that is correct within a factor of two or three.

Direct observation of absorptive transitions from the singlet ground state into triplet states is very difficult since the transitions are spin-forbidden and the corresponding extinction coefficient at least a factor of 10^6 lower than that in spin-allowed $\pi \rightarrow \pi^*$ transitions.

Proceeding finally from the two-electron wave function to the N-electron wave function, the following picture of the eigenstates of the dye

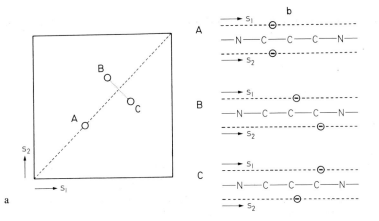

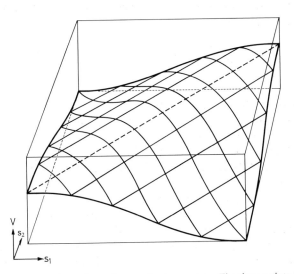

Fig. 1.13. a Three configurations of a two-electron system. The three points A, B, C in the cartesian coordinate system on the left belong to the three two-electron configurations pictured on the right. b Potential energy $V(s_1,s_2)$ of the two-electron system. (From KUHN, 1955)

molecule is obtained (Fig. 1.14). There is a ladder of singlet states $S_i (i = 1,2,3,...)$ containing also the ground state G. Somewhat displaced towards lower energies there is the ladder of triplet states $T_i (i = 1,2,3,...)$. The longest wavelength absorption is from G to S_1, the next absorption band from G to S_2, etc. By contrast the absorption from G to T_i is

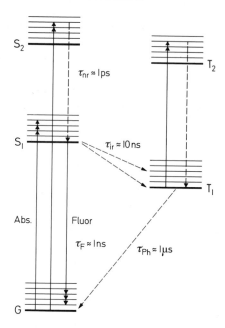

Fig. 1.14. Eigenstates of a typical dye molecule with radiative (solid lines) and non-radiative (broken lines) transitions

spin-forbidden. The absorptions $G - S_1$, $G - S_2$, etc., will usually have differing transition moments. The fate of a molecule after it has undergone an absorptive transition and finds itself in an excited state S_1 or S_2 is discussed in the next section.

1.3. Deactivation Pathways for Excited Molecules

There are very many processes by which an excited molecule can return directly or indirectly to the ground state. Some of these are schematically depicted in Fig. 1.14. It is the relative importance of these which mainly determines how useful a dye will prove in dye lasers. The one process that is directly used in dye lasers is the radiative transition from the first excited singlet state S_1 to the ground state G. If this emission, termed "fluorescence", occurs spontaneously, its radiative lifetime τ_{rf} is connected with the Einstein coefficient for spontaneous emission A and the oscillator strength f of the pertinent absorption band by

$$A \equiv 1/\tau_{rf} = (8\pi^2 \mu^2 e_0^2/m_0 c_0)\tilde{v}^2 f,$$

where μ is the refractive index of the solution, e_0 is the charge and m_0 is the mass of the electron, c_0 is the velocity of light, and $\tilde{v}$ denotes the wave number of the center of the absorption band. This relation is valid if the half-width of the emission band is small and its position is not shifted significantly from that of the absorption band. Since the f values of the transition are near unity in most dyes, the radiative lifetime τ_{rf} is typically of the order of a few nanoseconds. Generally, however, the fluorescence spectrum is broad and shows considerable Stokes shift. In this case the radiative lifetime can be computed with the aid of the following relation [STRICKLER and BERG, 1962]

$$\frac{1}{\tau_{rf}} = 2.88 \times 10^{-9} \mu^2 \frac{\int F(\tilde{v})d\tilde{v}}{\int \tilde{v}^{-3} F(\tilde{v})d\tilde{v}} \int_{\substack{\text{longest} \\ \text{wavelength} \\ \text{absorption band}}} \frac{\varepsilon(\tilde{v})d\tilde{v}}{\tilde{v}}$$

where $F(\tilde{v}) = dQ/d\tilde{v}$ is the fluorescence spectrum (in quanta Q per wave number) and $\varepsilon(\tilde{v})$ is the molar decadic extinction coefficient. There also exists the possibility that the energy hypersurface of the excited state will approach closely enough to the ground state for the molecule to tunnel through the barrier between them. It is then found in a very highly excited vibronic level of the ground state. This process is generally termed "internal conversion" and its rate constant is k_{SG}. There is very little theoretical knowledge about this and other related radiationless processes (ROBINSON and FROSCH, 1963). Experimental results (which range over several orders of magnitude of k_{SG}) will be discussed in detail in Chapter 4.

The internal conversion between S_2 (and higher excited states) and S_1 is usually extremely fast, taking place in less than 10^{-11} sec. This is the reason why fluorescence quantum spectra of dyes generally do not depend on the excitation wavelength. The only known exception to these general rules is the aromatic molecule azulene and its derivatives. The fluorescence of this molecule is directly from S_2 to the ground state, while internal conversion from S_1 to the ground state is so fast that no fluorescence originating in S_1 can be detected except with very high-peak power laser pulses (BIRKS, 1972).

The radiationless transition from an excited singlet state to a triplet state can be induced by internal perturbations (spin-orbit coupling, substituents containing nuclei with high atomic number) as well as by external perturbations (paramagnetic collision partners, like O_2 molecules in the solution or solvent molecules containing nuclei of high atomic number). These radiationless transitions are usually termed "intersystem crossing" and have the rate constant k_{ST}.

If some fluorescence quenching agent Q is present in a solution or gas mixture in a known concentration [Q], its contribution to the radiationless deactivation can be separated and expressed by a rate constant $k'_{SG} = k_Q \cdot [Q]$. Since for most quenchers quenching occurs at every encounter with an excited dye molecule, k_Q equals the diffusion-controlled bimolecular rate constant, given by $k_Q = 8RT/2000\eta$, where R is the gas constant, T the absolute temperature and η the viscosity of the solvent (OSBORNE and PORTER, 1965).

The quantum yield of fluorescence, ϕ_f, is then defined as the ratio of radiative and nonradiative transition rates

$$\phi_f = \frac{1/\tau_{rf}}{1/\tau_{rf} + k_{ST} + k_{SG} + k_Q \cdot [Q]}.$$

The denominator is the reciprocal of the observed fluorescence lifetime τ_f. The quantum yield of fluorescence can thus be calculated from the measured fluorescence lifetime τ_f and the radiative lifetime τ_{rf}, obtained through the absorption spectrum: $\phi_f = \tau_f/\tau_{rf}$. This method usually gives good results, whereas the direct experimental measurement of absolute quantum yields is very difficult (see Chapter 4).

The radiative transition from $T_1 \rightarrow G$ is termed "phosphorescence". In principle, one should be able to estimate the radiative lifetime τ_{rp} of the phosphorescence on the basis of the $G \rightarrow T_1$ absorption. Since this transition is normally completely obscured by absorption due to impurities, the phosphorescence radiative lifetime can be obtained from the quantum yield of phosphorescence, ϕ_p, and the observed phosphorescence lifetime τ_p, as for fluorescence: $\tau_p = \phi_p \tau_{rp}$. As expected for spin-forbidden transitions, it is extremely long, ranging from milliseconds to many seconds for molecules with small spin–orbit coupling. Consequently, even relatively slow quenching processes can lead to radiationless deactivation in liquid solution. Hence the observed τ_p is generally very low in liquid solution, becoming appreciable only at low temperatures, e.g. 77 °K in solid solutions. (A similar comment can be made regarding the long-lived emission resulting from the above-mentioned $\pi^* \leftarrow n$-transitions, which are usually very weak, if any fluorescence is observed at all.) Note that ϕ_p is the true quantum yield of phosphorescence. It is defined as the ratio of the phosphorescence rate to the total rate of radiative and radiationless deactivation $\phi_p = \dfrac{1/\tau_{rp}}{1/\tau_{rp} + k_{TG} + k_Q \cdot [Q]}$. By comparison, one often quotes the apparent quantum yield of phosphorescence ϕ_p^+, which relates the rate of phosphorescence emission to the rate of light absorption. The relation between the two is $\phi_p^+ = \phi_p \phi_T$. Here ϕ_T is the triplet formation efficiency, i.e. the ratio of the intersystem crossing rate k_{ST} to the total rate of deactivation of the fluorescent level:

$$\phi_T = \frac{k_{ST}}{1/\tau_{rf} + k_{ST} + k_{SG} + k_Q \cdot [Q]}.$$

Usually no measurable phosphorescence can be observed in liquid solutions. Here τ_p, which is identical with the triplet lifetime τ_T, is often determined by flash spectroscopy from the vanishing of the triplet–triplet absorption bands. Assuming the value of the radiative lifetime at elevated temperatures is the same as at low temperatures, the quantum yield ϕ_p in liquid solutions can be obtained. To give a typical example, for the eosin-di-anion in methanol (7×10^{-5} mole/l) the phosphorescence life-time at room temperature is $\tau_p = 1.5$ msec, the apparent quantum yield of phosphorescence $\phi_p^+ = 3.9 \times 10^{-3}$, the triplet quantum yield $\phi_T = 0.45$, hence the radiative lifetime $\tau_{rp} = 0.2$ sec (PARKER and HATCHARD, 1961).

The lowest singlet and triplet states can also be deactivated by a long-range radiationless energy transfer to some other dye molecule. For this to happen, the absorption band of the latter, "acceptor" molecule must overlap the fluorescence or phosphorescence band of the former, "sensitizer" molecule. The so-called critical distance, at which the quantum yield is reduced by a factor of two, can reach values of 10 nm for the case of singlet–singlet energy transfer. For details of these and some other processes resulting in "delayed fluorescence", which is less important in the present context, the reader is referred to PARKER (1968).

Other deactivation pathways for the excited molecule include reversible and irreversible photochemical reactions, e.g. protolytic reactions, cis-trans-isomerizations, dimerizations, ring-opening reactions and many others. Some of these which are of major importance in dye lasers are discussed in Chapter 4.

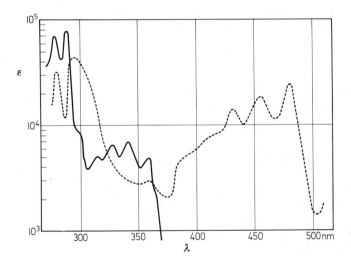

Fig. 1.15. Triplet-triplet absorption spectrum (broken line) of 1,2-benzanthracene in hexane (solid line: absorption spectrum from ground to excited states). (From LABHART, 1964)

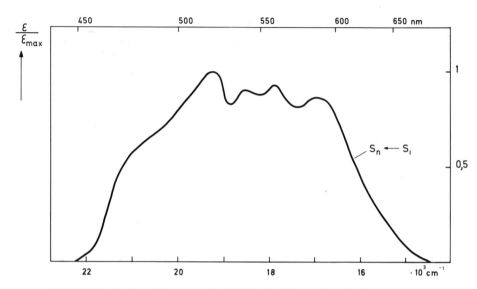

Fig. 1.16. $S_1 \rightarrow S_n$-absorption spectrum of 1,2-benzanthracene in ethanol. (From MÜLLER and SOMMER, 1969)

The absorptive transitions of excited molecules into higher excited states can constitute a loss mechanism for the pump as well as the dye-laser radiation. It thus is of utmost importance for dye-laser operation.

Triplet–triplet absorption is measured by flash photolysis as well as by photostationary methods (PORTER and WINDSOR, 1958; LABHART, 1964; ZANKER and MIETHKE, 1957b). An example of dye spectra thus obtained is given in Fig. 1.15. It is much more difficult to measure the absorption from the short-lived excited singlet state and has only become possible through the use of laser spectroscopy methods (NOVAK and WINDSOR, 1967; MÜLLER, 1968; BONNEAU et al., 1968). So far spectra for only about 20 molecules have been recorded (BIRKS, 1970). An example is given in Fig. 1.16.

1.4. Laser-Pumped Dye Lasers

1.4.1. Oscillation Condition

From the foregoing discussion of the spectroscopic properties of dyes one might come to the conclusion that there are two possible ways, at least in principle, of using an organic solution as the active medium in a laser: One might utilize either the fluorescence or the phosphorescence emission. At first sight the long lifetime of the triplet state makes phosphorescence look more attractive. On the other hand, due to the strongly forbidden transition, a very high concentration of the active species is required to obtain an amplification factor large enough to overcome the inevitable cavity losses. In fact, for many dyes this concentration would be higher than the solubility of the dyes in any solvent. A further unfavorable property of these systems is that there will almost certainly be losses due to triplet–triplet absorption. It must be remembered that triplet–triplet absorption bands are generally very broad and diffuse and the probability they will overlap the phosphorescence band is high. Because of these difficulties no laser using the phosphorescence of a dye has yet been reported (a preliminary report [MORANTZ, 1962 and 1963] was evidently in error). The possibility cannot be excluded, however, that further study of phosphorescence and triplet–triplet absorption in molecules of different types of chemical constitution might eventually lead to a laser operating, for example, at the temperature of liquid nitrogen. On the other hand, the probability for this seems low at present and these systems will not be considered here. For a more detailed discussion of phosphorescent systems the reader is referred to LEMPICKI and SAMELSON [1966].

If the fluorescence band of a dye solution is utilized in a dye laser, the allowed transition from the lowest vibronic level of the first excited singlet state to some higher vibronic level of the ground state will give a high amplification factor even at low dye concentrations. The main complication in these systems is the existence of the lower-lying triplet states. The intersystem crossing rate to the lowest triplet state is high enough in most molecules to reduce the quantum yield of fluorescence to values substantially below unity. This has a twofold consequence: Firstly, it reduces the population of the excited singlet state, and hence the amplification factor; and secondly, it enhances the triplet–triplet absorption losses by increasing the population of the lowest triplet state. Assume a light-flux density which slowly rises to a level P [quanta $\sec^{-1} \operatorname{cm}^{-2}$], a total molecular absorbing cross-section σ [cm^2][1], a quantum yield ϕ_T of triplet formation, a triplet lifetime τ_T, populations of the triplet and ground state of n_T and n_0 [cm^{-3}], respectively, and, neglecting the small population of the excited singlet state, a total concentration of dye molecules of $n = n_0 + n_T$. A steady state is reached when the rate of triplet formation equals the rate of deactivation:

$$P\sigma n_0 \phi_T = n_T/\tau_T . \qquad (1.1)$$

Thus, the fraction of molecules in the triplet state is given by

$$n_T/n = P\sigma\phi_T\tau_T/(1 + P\sigma\phi_T\tau_T) . \qquad (1.2)$$

---

[1] The absorption cross-section σ can be determined from the molar decadic extinction coefficient ε by $\sigma = 0.385 \times 10^{-20}\varepsilon$. Here σ is given in cm^2, if ε is measured in liter/(mole · cm).

Assuming some typical values for a dye, $\sigma = 10^{-16}$ cm^2, $\phi_T = 0.1$ (corresponding to a 90% quantum yield of fluorescence), and $\tau_T = 10^{-4}$ sec, the power to maintain half of the molecules in the triplet state is $P_{1/2} = 10^{21}$ quanta sec^{-1} cm^{-2}, or an irradiation of only 1/2 kW cm^{-2} in the visible part of the spectrum. This is much less than the threshold pump power calculated below. Hence a slowly rising pump light pulse would transfer most of the molecules to the triplet state and deplete the ground state correspondingly. On the other hand, the population of the triplet level can be held arbitrarily small, if the pumping light flux density rises fast enough, i.e. if it reaches threshold in a time t_r which is small compared to the reciprocal of the intersystem crossing rate $t_r \ll 1/k_{ST}$. Here t_r is the risetime of the pump light power, during which it rises from zero to the threshold level. For a typical value of $k_{ST} = 10^7$ sec^{-1}, the risetime should be less than 100 ns. This is easily achieved, for example with a giant-pulse laser as pump light source, since giant pulses usually have risetimes of 5–20 nsec. In such laser-pumped dye-laser systems one may neglect all triplet effects in a first approximation.

We thus can restrict our discussion to the singlet states. Molecules that take part in dye-laser operation have to fulfill the following cycle (Fig. 1.17): Absorption of pump radiation at $\tilde{v}_p$ and with cross-section σ_p lifts the molecule from the ground state with population n_0 into a higher vibronic level of the first (or second) excited singlet state S$_1$ (or S$_2$) with a population n_1' (or n_2'). Since the radiationless deactivation to the lowest

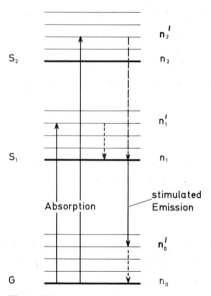

Fig. 1.17. Pump cycle of dye molecules

level of S_1 is so fast (see Section 1.3), the steady-state population n_1' is negligibly small, provided the temperature is not so high that this vibronic level is already thermally populated by the Boltzmann distribution of the molecules in S_1. At room temperature $kT = 200\ \text{cm}^{-1}$, so that this is not the case. Stimulated emission then occurs from the lowest vibronic level of S_1 to higher vibronic levels of G. Again the population n_0' of this vibronic level is negligible since the molecules quickly relax to the lowest vibronic levels of G.

It is easy then to write down the oscillation condition for a dye laser (SCHÄFER, 1968; SCHÄFER et al. 1968). In its simplest form a dye laser consists of a cuvette of length $L\,[\text{cm}]$ with dye solution of concentration $n\,[\text{cm}^{-3}]$ and of two parallel end windows carrying a reflective layer each of reflectivity R for the laser resonator. With n_1 molecules/cm^3 excited to the first singlet state, the dye laser will start oscillating at a wave number $\tilde{v}$, if the overall gain is equal to or greater than one:

$$\exp\left[-\sigma_a(\tilde{v})n_0 L\right] R \exp\left[+\sigma_f(\tilde{v})n_1 L\right] \geqq 1. \tag{1.3}$$

Here $\sigma_a(\tilde{v})$ and $\sigma_f(\tilde{v})$ are the cross-sections for absorption and stimulated fluorescence at $\tilde{v}$, respectively, and n_0 is the population of the ground state. The first exponential term gives the attenuation due to reabsorption of the fluorescence by the long-wavelength tail of the absorption band. The attenuation becomes the more important, the greater the overlap between the absorption and fluorescence bands. The cross-section for stimulated fluorescence is related to the Einstein coefficient B

$$\sigma_f(\tilde{v}) = g(\tilde{v})Bh\tilde{v}/c_0 \underset{\substack{\text{fluorescence}\\\text{band}}}{\int} g(\tilde{v})\mathrm{d}\tilde{v} = 1. \tag{1.4}$$

Substituting the Einstein coefficient A for spontaneous emission according to

$$B = \frac{1}{8\pi\tilde{v}^2}A\frac{1}{h\tilde{v}} \tag{1.5}$$

and realizing that $g(\tilde{v})A \cdot \phi_f = Q(\tilde{v})$, the number of fluorescence quanta per wave number interval, one obtains

$$\sigma_f(\tilde{v}) = \frac{1}{8\pi c_0\tilde{v}^2} \cdot \frac{Q(\tilde{v})}{\phi_f}. \tag{1.6}$$

Since the fluorescence band usually is a mirror image of the absorption band, the maximum values of the cross-sections in absorption and emission are found to be equal:

$$\sigma_{f,\text{max}} = \sigma_{a,\text{max}}. \tag{1.7}$$

Taking the logarithm of (1.3) and rearranging it leads to a form of the oscillation condition which makes it easier to discuss the influence of the various parameters:

$$\frac{S/n + \sigma_a(\tilde{v})}{\sigma_f(\tilde{v}) + \sigma_a(\tilde{v})} \leqq \gamma(\tilde{v}), \tag{1.8}$$

where $S = (1/L)\ln(1/R)$ and $\gamma(\tilde{v}) = n_1/n$.

The constant S on the left-hand side of (1.8) only contains parameters of the resonator, i.e. the active length L, and reflectivity R. Other types of losses, like scattering, diffraction, etc., may be accounted for by an effective reflectivity, R_{eff}. The value $\gamma(\tilde{v})$ is the minimum fraction of the molecules that must be raised to the first singlet state to reach the threshold of oscillation. One may then calculate the function $\gamma(\tilde{v})$ from the absorption and fluorescence spectra for any concentration n of the dye and value S of the cavity. In this way one finds the frequency for the minimum of this function. This frequency can also be obtained by differentiating (1.8) and setting $d\gamma(\tilde{v})/d\tilde{v} = 0$. This yields

$$\frac{\sigma_a'(\tilde{v})}{\sigma_f'(\tilde{v}) + \sigma_a'(\tilde{v})}\left(\sigma_f(\tilde{v}) + \sigma_a(\tilde{v})\right) = S/n \tag{1.9}$$

(prime means differentiation with respect to $\tilde{v}$) from which the start-oscillation frequency can be obtained.

Figure 1.18 represents a plot of the laser wavelength λ (i.e. the wavelength at which the minimum of $\gamma(\tilde{v})$ occurs) versus the concentration with S a fixed parameter. Similarly, Fig. 1.19 shows the laser wavelength versus the active length L of the cuvette with dye solution, with the concentration of the dye as a parameter. Both figures apply to the dye 3,3'-diethylthiatricarbocyanine (DTTC). These diagrams demonstrate the wide tunability range of dye lasers which can be induced by changing the concentration of the dye solution, or length, or Q of the resonating cavity. They also show the high gain which permits the use of extremely small active lengths.

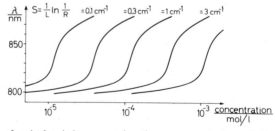

Fig. 1.18. Plot of calculated laser wavelength vs. concentration of the laser dye 3,3'-diethylthiatricarbocyanine bromide, with S as a parameter. (From Schäfer, 1968)

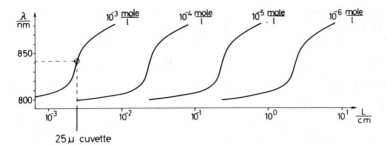

Fig. 1.19. Plot of calculated laser wavelength vs. active length of the laser cuvette, with the concentration of the solution of the dye 3,3'-diethylthiatricarbocyanine bromide as a parameter. (From SCHÄFER, 1968)

The absorbed power density W necessary to maintain a fraction γ of the molecular concentration n in the excited state is

$$W = \gamma n h c_0 \tilde{v}_p / \tau_f , \tag{1.10}$$

and the power flux, assuming the incident radiation is completely absorbed in the dye sample,

$$P = W/n\sigma = \gamma h c_0 \tilde{v}_p / \tau_f \sigma \tag{1.11}$$

where $\tilde{v}_p$ is the wave number of the absorbed pump radiation. If the radiation is not completely absorbed, the relation between the incident power W_{in} and the absorbed power is $W = W_{in}[1 - \exp(-\sigma_p n_0 L)]$. Since in most cases $n \approx n_0$, this reduces for optically thin samples to $W = W_{in} \sigma_p n L$. The threshold incident power flux, P_{in}, then is

$$P_{in} = (\gamma h c_0 \tilde{v}_p)/(\tau_f n \sigma^2 L) .$$

In the above derivation of the oscillation condition and concentration dependence of the laser wavelength, broad-band reflectors have been assumed. The extension to the case of wavelength-selective reflectors and/or dispersive elements in the cavity is straightforward and will not be treated here.

1.4.2. Practical Pumping Arrangements

A simple structure for dye-solution laser was used in many of the early investigations. It is still useful for exploratory studies of new dyes. It consists of a square spectrophotometer cuvette filled with the dye solution (Fig. 1.20) which is excited by the beam from a suitable laser. As shown in the figure, the resonator is formed by the two glass-air interfaces of the polished sides of the cuvette (heavy lines in Fig. 1.20a)

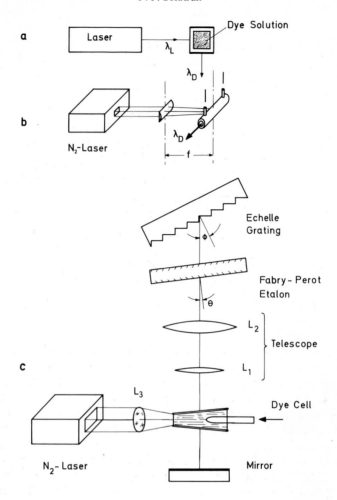

Fig. 1.20. Resonator configurations for laser-pumped dye lasers. (λ_L pumping-laser beam, λ_D dye-laser beam, AR: anti-reflection coating, f focal length of cylindrical lens

and the exciting laser and dye-laser beams are at right angles. Reflective coatings on the windows consisting of suitable metallic or multiple dielectric layers enhance the Q of the resonator. It is also possible to use antireflective coatings or Brewster windows at the cuvette and separate resonator mirrors. In the transverse pumping arrangement (Fig. 1.20a) the population inversion in the dye solution is nonuniform along the exciting laser beam, since the exciting beam is attenuated in the solution. Consequently, at a higher concentration threshold might only be

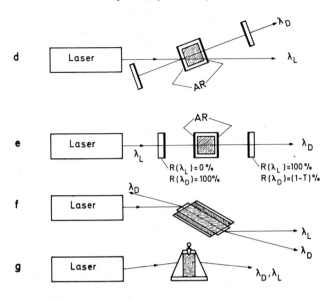

Fig. 1.20 d–g.

reached in a thin layer directly behind the entrance window of the exciting beam. This gives rise to large diffraction losses and beam divergence angles. If the concentration is not too high, say, less than 10^{-4} molar, the measured threshold pump power in such an arrangement can be used to test the value calculated from the oscillation condition. Using (1.8) and (1.11), known spectral data, and the observed spontaneous decay time of fluorescence $\tau_f = 2$ nsec, one computes a threshold pump power flux of $P = 1.2 \times 10^5$ W/cm² for the exciting giant pulse ruby laser and a 10^{-4} molar solution of 3,3'-diethylthiatricarbocyanine bromide in methanol in a 1 cm square spectrophotometric cuvette. The experimental value is 3.0×10^5 W/cm², while the calculated and observed values for a 10^{-3} molar solution are 1.2×10^4 W/cm² and $1.2 \cdot 10^5$ W/cm², respectively (VOLZE, 1969). The higher value of the observed threshold for the 10^{-4} molar solution is easily accounted for by neglecting of triplet losses. The risetime of the exciting pulse was 8 nsec in these experiments, while the spontaneous lifetime of the dye is 2 nsec. An estimate of the quantum yield of fluorescence gives $\phi_f = 0.4$. Hence, an appreciable fraction of the molecules has already been converted to the triplet state at threshold. The larger discrepancy between calculated and observed values for the 10^{-3} molar solution can be traced to the high diffraction losses. In this experiment the exciting laser beam is attenuated to $1/e$ of incident intensity in a depth of only 85 μm. A similar transverse arrangement is often used for nitrogen laser-pumped dye lasers (Fig. 1. 20 b). Here the emission of a nitrogen laser is focused by a cylindrical lens into a line that coincides with the axis of a quartz capillary of inside diameter $d \approx 1$ mm. The transmitted pump radiation is reflected by an aluminum coating at the back surface of the capillary tube. If the beam divergence of the nitrogen laser is α and the length of the cylindrical lens f, then the width of the focal line is $H = f \cdot \alpha$ and is best so chosen that H is about one fourth of the inside diameter of the tube. The dye concentration c is then adjusted so that the absorption $A = 2\varepsilon \cdot c \cdot d \approx 1$. The endfaces of the tube can either be normal to the axis and act directly as resonator mirrors, or set at the Brewster angle for use with external mirrors.

Another arrangement developed for the pumping of dye lasers with a nitrogen laser is shown in Fig. 1.20 c (Hänsch, 1972). The dye cell is 10 mm long and is made of 12-mm diameter Pyrex tubing. Antireflection-coated quartz windows are sealed to the ends of the cell under a wedge angle of about 10 degrees to avoid internal cavity effects. The dye solution is transversely circulated by means of a small centrifugal pump at a flow speed of about 1 m/sec in the active region. The nitrogen laser emission is focused by a spherical lens of 135 mm focal length into a line of about 0.15 mm width at the inner cell wall. To provide a near-circular active cross-section, the dye concentration is adjusted so that the penetration of the pump light is also of the order of 0.15 mm. A plane outcoupling mirror is mounted at one end of the optical cavity at a distance of 50 mm. Because of the large ratio of this distance to the diameter of the active region, the latter acts as an active pinhole and the emerging beam is essentially diffraction-limited and has a beam divergence (at 600 nm emission wavelength) of 2.4 mrad. The other end of the cavity contains an inverted telescope, consisting of two lenses L_1 and L_2 of focal lengths $f_1 = 8.5$ mm and $f_2 = 185$ mm (antireflectioncoated multielement systems corrected for spherical aberration and coma within $\lambda/8$) and a separation between cell and lens L_1 of 75 mm. The lenses are used slightly off-axis to avoid back reflection and etalon effects. The initial waist size of 0.08 mm is thus enlarged to 4 mm and the beam divergence is reduced to 0.05 mrad. This is especially important if a diffraction grating in Littrow mounting is to be used instead of a second mirror. In this case, if the beam were unexpanded, it would eventually burn the grating and the spectral resolution would be very low.

A better configuration for laser-pumped dye lasers is the longitudinal arrangement (Sorokin et al., 1966 b) of Fig. 1.20 e. In this arrangement the exciting laser beam passes through one of the resonator mirrors of the dye laser. For best effect this is coated by a dielectric multilayer mirror with a very low reflection coefficient at the exciting-laser wavelength and a high one at the dye-laser wavelength. Concentration and depth of the solution should be adjusted so that the excitation power is very nearly uniform over the whole volume of the cuvette. This results in a much lower beam divergence. Typical values of $3-5$ mrad were reported. It is often more convenient to orient the exciting-laser beam a few degrees to the normal of the mirror for spatial separation of the exciting-laser and the dye-laser beams. If the dimension of the cuvette is much smaller than the cavity length, the exciting-laser beam may pass by the side of the laser mirror (Fig. 1.20d). This eliminates the requirement for a special mirror system that must withstand the full power of the exciting laser.

Another advantage of the longitudinal pumping arrangement is the possibility of using an extremely small depth of dye solution. VOLZE (1969) described a dye laser consisting of two mirrors in direct contact with the dye solution and kept apart by spacers ranging from 100 µm down to 5 µm. The mirrors had 80% transmission at the exciting ruby-laser wavelength, and a constant 98% reflection over the range of the dye-laser emission. Using a 10^{-3} molar solution of 3,3'-diethylthiatricarbocyanine bromide (or other anion), the observed threshold for a 5 µm spacer is 350 kW/cm^2, while the computed value is 120 kW/cm^2. This is about the same error factor as in the above transverse pumped case with the 10^{-4} molar solution. Here, too, the error should be due mainly to neglect of the triplet losses. The operation of a dye laser of only 5 µm width demonstrates the extremely high gain in the dye solution. This experiment indicates an amplification factor of 1.02 per 5 µm which implies a gain of $G = 170$ dB/cm.

Superradiant dye lasers can be pumped either by a normal giant-pulse laser or by ultrashort pulses. In the first case (Fig. 1.20 f), the exciting-laser pulse passes through a cuvette of small optical length compared to the length of the exciting pulse.

Typically, the solution is contained in a 5-cm Brewster cuvette with quartz tubing and windows (VOLZE, 1969). The solvent DMSO has the same refractive index ($n = 1.48$) as the cuvette walls. The cuvette tubing is surrounded by a cylinder containing a suitable dye solution in a solvent of higher refractive index so that fluorescence light incident on the cuvette walls is absorbed. This eliminates the unwanted closed-feedback path. Near the peak of the exciting-laser pulse the spatial distribution of pump light in the cuvette is nearly stationary. With this assumption, the amplification can be computed for spontaneously emitted fluorescence quanta that originate at one window and travel along the optical axis towards the other window. As expected, in this model the superradiant dye-laser intensities are almost equal in both forward and backward directions, and the beam divergence is determined by the aspect ratio of the cuvette. A plot of the superradiant versus the exciting intensity for this

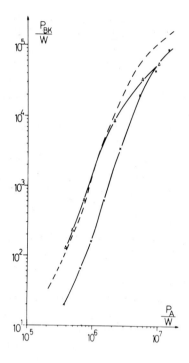

Fig. 1.21. Superradiant power P_{BK} from a Brewster angle cuvette vs. pumping laser power P_A for a 3×10^{-5} M solution of 3,3'-diethylthiacarbocyanine bromide. (Triangles: emission in backward direction; crosses: emission in forward direction. The broken line gives the theoretical expectation). (From Volze, 1969)

configuration with a 3×10^{-5} molar solution of DTTC is shown in Fig. 1.21.

For traveling-wave excitation of a superradiant dye laser, the solution is contained in a wedged (10°) cell typically 2 cm in depth (Mack, 1969). A methanol solution of DTTC or some similar dye is pumped by a ruby-laser pulse of a few psec half-width and 5 GW peak power. This results in an almost complete inversion of the dye solution at the position of the exciting pulse with correspondingly high gain. For a pulse width of 5 psec, corresponding to an inverted region of 2 mm, spontaneously emitted fluorescence quanta passing through this region would experience a maximum gain of 50 dB in a 10^{-4} molar solution of DTTC. It is assumed that no saturation effects occur and that the relaxation time from the initial Franck-Condon state to the upper laser level is negligible compared to the duration of the exciting pulse. Since the vibrational relaxation time of dyes in solution is generally of the order of a few psec, and since saturation certainly must occur as the exciting pulse travels through the cuvette, the actual gain is less than the above value. The traveling-wave nature of the superradiant dye-laser emission may be verified by measuring the forward to backward ratio of the dye-laser emission. It turned out to be 100 to 1 in this experiment, and the beam divergence was 15 mrad.

The polarization of the dye laser beam in the longitudinal and transverse arrangements is determined mainly by the polarization of the exciting laser beam, the relative orientation of the transition moments in the dye molecule for the pumping and laser transitions, and the rotational diffusion–relaxation time. The latter is determined by solvent viscosity, temperature and molecular size. The direction of the transition moments of the fluorescence and the long-wavelength absorption is identical, since the same electronic transition is involved in both processes. For rotational diffusion–relaxation times which are long compared to the exciting pulse, the following table gives the theoretically expected relative orientation of the pumping and dye-laser polarization. In two cases one obtains a depolarized dye-laser beam. There the emissive transition moments of the excited molecules are directed along the resonator axis. Hence the molecules must first rotate before they can radiate along the resonator axis. Since this rotation is effected by diffusion, the resulting angular orientation distribution is essentially isotropic and the laser output is depolarized. The same is true when the rotational diffusion-relaxation time is short compared to the pumping pulse. A special case is that of large, ring-type molecules like the phthalocyanines. These have two degenerate transitions with orthogonal transition moments in the plane of the ring. The polarization of dye lasers with these compounds is given in brackets in the following table. Very few such possibilities have been experimentally investigated and verified (SOROKIN et al., 1967; SEVCHENKO et al., 1968; McFARLAND, 1967). In addition, the polarization of dye laser beams can, of course, be manipulated in obvious ways by introducing into the resonator polarizing elements like Brewster windows.

Relative orientation of pumping and dye-laser polarizations	Relative orientation of transition moments in absorption and emission	
	‖ to each other	⊥ to each other
Longitudinal pumping	‖ (depol.)	⊥ (depol.)
Transverse pumping, pump laser polar. ⊥ to P	‖ (depol.)	⊥ (⊥)
Transverse pumping, pump laser polar. ‖ to P	depol. (⊥)	depol. (depol.)

P = plane containing exciting and dye-laser beams. Data in brackets refer to large, ring-type molecules, e.g. phthalocyanines.

The above techniques employ cells for the containment of liquid dye solutions, but even simpler arrangements are possible with solid solutions of dyes or dye crystals. Various dyes have been incorporated in polymethylmethacrylate and other plastics (SOFFER and McFARLAND, 1967).

If two parallel faces of such a sample are optically polished, the Fresnel reflection of the air-plastics interface can be utilized as the air-glass reflection in the spectrophotometer cuvette of Fig. 1.20a. A number of dyes have been dissolved in gelatin. The molten gelatin was poured onto a glass slide in a layer a few microns thick. After hardening, this thin film could be stripped off the glass slide and showed strong superradiant emission when the emission of a nitrogen laser was focused on it with a cylindrical lens (HÄNSCH et al., 1971).

Organic single crystals, e.g. fluorene crystals containing $2 \cdot 10^{-3}$ anthracene, which grow and cleave well, can also be pumped by a nitrogen laser, using the two cleaved faces of the laser crystal itself as the optical resonator (KARL, 1972). In the case of the fluorene crystal containing anthracene as laser dye, the crystals had a thickness of $0.3 - 0.7$ mm and absorbed the pump light almost completely. Laser emission was found at 408 nm.

No laser emission from dyes in the vapor phase has yet been reported. It is to be expected, however, that this will be achieved in the near future using a high-power nitrogen laser and suitable dyes.

Pump lasers used so far with one or the other of the above arrangements include neodymium (1.06 μm) and its harmonics (532 nm, 354 nm, and 266 nm), ruby (694 nm) and its harmonic (347 nm), nitrogen (337 nm), and other dye lasers.

If the repetition rate of the pump pulses is increased above about one shot per minute, the heat generated via radiationless transitions in the dye molecules and energy transfer to the solvent can cause schlieren in the cuvette which reduce the dye-laser output. It is thus advisable to use a flow system so that the cuvette contains fresh solution for every shot. Obviously, this is also the way to deal with problems of photochemical instability. However, these rarely arise in laser-pumped dye lasers, except with some very unstable infrared sensitizing dyes.

1.4.3. Time Behavior and Spectra

Most often the pump pulse is of approximately Gaussian shape and its full width at half maximum power is less than the reciprocal of the intersystem-crossing rate constant k_{ST}. Thus, we can expect the following time behavior, neglecting finer detail for the moment. Shortly after the pump pulse reaches threshold level, dye-laser emission starts. The dye-laser output power closely follows the pump power till it drops below threshold, when dye-laser emission stops. The dye-laser pulse shape should thus closely resemble that of the part of the pump pulse above the threshold level. This behavior was indeed observed by SCHÄFER et al.

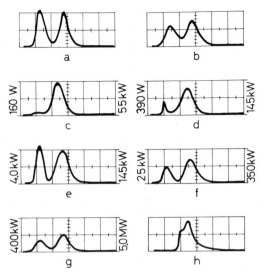

Fig. 1.22. Oscillograms of fluorescence, dye-laser and pumping-laser-pulses. (Sweep speed: 10 nsec/division for a–g, and 2 nsec/div. for h). a Pumping pulse and delayed reference pulse, b spontaneous fluorescence of a 10^{-5} M solution of the dye 3,3'-diethylthiatricarbocyanine bromide in methanol and delayed reference pulse, c–f dye-laser pulse from a 10^{-3} M solution of the dye and delayed reference pulse. The power corresponding to two major scale divisions is annotated to the right of each oscillogram for the pumping ruby laser pulse, to the left for the dye-laser pulse. g Dye-laser pulse just above threshold. (From SCHÄFER et al., 1966)

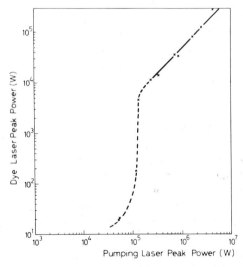

Fig. 1.23. Dye-laser peak power vs. pumping-laser peak power for a 10^{-3} M solution of 3,3'-diethylthiatricarbocyanine bromide in methanol. (From SCHÄFER et al., 1966)

(1966) pumping transversely a methanol solution of the dye DTTC in a square spectrophotometer cuvette of 1×1 cm dimension, polished on all sides. Figure 1.22 shows oscillograms of this pump and dye-laser pulse at different peak pump power levels. From an evaluation of these oscillograms the input–output plot shown in Fig. 1.23 is obtained. It clearly shows how the dye fluorescence intercepted by the photocell is increased by several orders of magnitude from the spontaneous level, while the pump power is increased slightly above threshold. When the threshold is reached at the foot of the pump pulse, so that pump and dye-laser pulse forms are practically identical, the expected linear relationship between input and output obtains. The efficiency in this case was found to be 10%.

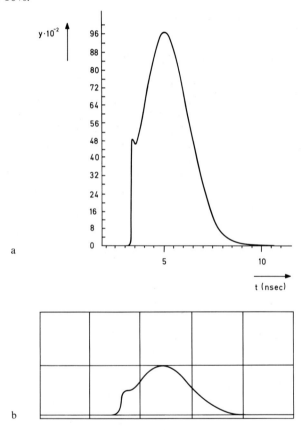

Fig. 1.24. Comparison of a computer solution of the rate equations and an actual oscillogram. a Computer solution for $N_{pump}/N_t = 350$, $\tau/t_c = 3.0$, $T_1/t_c = 6.0$, $y = Q/(t \cdot W_{max})$; b oscillogram of a chloro-aluminium-phthalocyanine laser. (Mirror spacing: 5 cm, mirror reflectivities: 80% and 98%, sweep speed: 20 nsec/division) (From SOROKIN et al., 1967)

A more detailed treatment of the time behavior of the laser-pumped dye laser was given by SOROKIN et al (1967) who solved the rate equations for the excited state population in the dye and the photons in the cavity

$$d N_1/dt = W(t) - (N_1/N_t) \cdot (Q/t_c) - N_1/\tau_f$$
$$dQ/dt = (Q/t_c) \cdot (N_1/N_t - 1).$$

The quantities appearing in these equations are as follows: N_1 is the excited-state population, N_t is the threshold inversion, Q is the number of quanta in the cavity, t_c is the resonator lifetime, τ_f is the fluorescence lifetime, and $W(t)$ denotes the pumping pulse which was assumed to have a Gaussian distribution with half-width at half-power points equal to T_1, i.e.

$$W(t) = W_{max} \cdot \exp\left[-(t\sqrt{\ln 2}/T_1)^2 \right]$$

and normalized to $\int_{-\infty}^{+\infty} W(t)dt = N_{pump}$, the total number of pumping photons. Digital computer solutions were obtained for a large number of parameter combinations, many of which agreed well with the experimentally observed time behavior even in such details as risetime and initial overshoot of the dye-laser pulse. A sample computer solution and an oscillogram of the dye-laser output for one set of parameters are reproduced from this work in Fig. 1.24.

As explained above, it is not difficult to calculate the oscillation wavelength of a laser-pumped dye laser near threshold. Figure 1.25 gives the

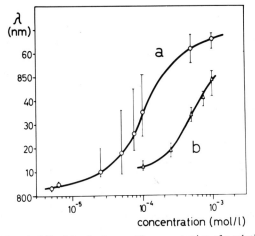

Fig. 1.25. Wavelength shift of the dye laser with concentration of a solution of 3,3'-diethylthiatricarbocyanine bromide in methanol, curve a for a silvered and curve b for an unsilvered cuvette. The circles and triangles give the wavelengths of the intensity maxima, the bars show the bandwidth of the laser emission. (From SCHÄFER et al., 1966)

observed dye-laser wavelength as a function of DTTC dye concentration for a low and a high value of cavity Q (SCHÄFER et al., 1966). These curves show satisfactory agreement with the computed curves of Fig. 1.18. The time-integrated spectral width of the dye-laser emission is indicated in Fig. 1.25 by bars. The width is measured with a constant pump power of 5 MW. It is seen to be rather broad in the middle of the concentration range for the high Q curve. This is due to the simultaneous and/or consecutive excitation of many longitudinal and transverse modes, which is also manifested by the increase in beam divergence. Several authors (FARMER et al., 1968; BASS and STEINFELD, 1968; GIBBS and KELLOG, 1968; VREHEN, 1971) have studied the wavelength sweep of the dye-laser emission. The results obtained are not easy to interpret. Frequently, there is a sweep from short to long wavelength of up to 6 nm, in other cases the sweep direction is reversed towards short wavelengths after reaching a maximum wavelength. Probably part of this sweep can be traced to triplet–triplet absorption, which should increase with time as the population density grows in the triplet level. Another part might be attributable to thermally induced phase gratings and similar disturbances of the optical quality of the resonator. Still another contributing factor might be the relatively slow rotational diffusion of large dye molecules in solvents like DMSO or glycerol. Inhomogeneous broadening of the ground state, on the other hand, although suggested as an explanation (BASS and STEINFELD, 1968), may be excluded on this time scale as a reason for the observed sweep. This explanation assumes that vibrational excitation of the ground state of the dye molecule has a lifetime of at least a few nsec in the solution but, in fact, it is deactivated in a few psec through collisions with the solvent molecules.

The dependence of the laser wavelength on the resonator Q, i.e. mainly on cell length and mirror reflectivities, can also be deduced from the start oscillation condition as described in subsection 1.4.1. The variation with Q was first noted by SCHÄFER et al. (1966) when they put silver reflectors on the faces of the dye cell. This is shown in Fig. 1.25. The dependence on the cell length was experimentally observed by FARMER et al. (1969).

The temperature dependence of the dye-laser wavelength can be similarly explained. SCHAPPERT et al. [1968] found that the laser emission of DTTC in ethanol shifted towards shorter wavelengths with decreasing temperature. This is caused by the narrowing of the fluorescence and the absorption band with decreasing temperature which results in higher gain and fewer reabsorption losses near the fluorescence peak. The solvent has a most important influence on the wavelength and efficiency of the dye-laser emission.

The first observation of large shifts of the dye-laser emission with changing solvents is reproduced in the following table for a 10^{-4} molar solution of the dye 1,1'-diethyl-γ-nitro-4,4'-dicarbocyanine tetrafluoroborate (SCHÄFER et al., 1966):

Solvent	Laser wavelength (in nm)
Methanol	796
Ethanol	805
Acetone	814
Dimethylformamide	815
Pyridine	821
Benzonitrile	822

In this case the nitro group, as a highly polar substituent in the middle of the polymethine chain, interacts with the dipoles of the solvent to shift the energy levels of the dye. This type of solvent shift is especially large for dyes whose dipole moments differ appreciably in the ground and excited states. The transition from ground to excited state by light absorption is fast compared with the dipolar relaxation of the solvent molecules. Hence the dye molecule finds itself in a nonequilibrium Franck-Condon state following light absorption, and it relaxes to an excited equilibrium state within about 10^{-13} to 10^{-11} sec. Similarly, the return to the equilibrium ground state is also via a Franck-Condon state, followed by dipolar relaxation:

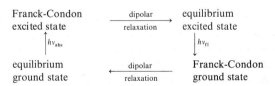

Consider the typical example of p-dimethylaminonitrostilbene. It has a dipole moment of 7.6 debye in the ground state and 32 debye in the excited state (LIPPERT, 1957). Therefore, it is clear that in a polar solvent like methanol the equilibrium excited state is lowered even more strongly than the equilibrium ground state by dipole–dipole interaction. In the present case, for example, the absorption maximum occurs at 431 nm and the fluorescence maximum at 757 nm in i-butanol, while the corresponding values for cyclohexane are 415 nm and 498 nm, respectively. Similar shifts are found in laser emission.

Gronau et al. [1972] report laser emission for the similar case of the dye 2-amino-7-nitrofluorene dissolved in 1,2-dichlorobenzene.

Another property of the solvent that can have an important influence on the dye-laser emission is the acidity of the solvent relative to that of the dye. As already mentioned in Section 1.2, many dyes show fluorescence as cations, neutral molecules, and anions. Correspondingly, the dye laser emission of such molecules usually changes with the pH of the solution, since generally the different ionization states of the molecule fluoresce at different wavelengths.

An important subdivision of these dyes is that of molecules, whose acidity in the excited state is considerably different from that in the ground state due to changes of the π-electron distribution with excitation. This might cause the molecule to lase or take up a proton from the solvent. After return to the ground state, the reverse action occurs, leaving the original molecule in the ground state. Schematically the reaction is

$$(AH)^* \xrightarrow[+\overleftarrow{H}^{\oplus}]{-H^{\oplus}} (A^{\ominus})^* \qquad\qquad A^* \xrightarrow[-\overleftarrow{H}^{\ominus}]{+H^{\oplus}} (AH^{\oplus})^*$$

$$\uparrow h\nu_{abs} \qquad \downarrow h\nu_{fl} \qquad\qquad \text{or} \qquad \uparrow h\nu_{abs} \qquad \downarrow h\nu_{fl}$$

$$AH \xleftarrow[-\overrightarrow{H}^{\oplus}]{+H^{\oplus}} A^{\ominus} \qquad\qquad A \xleftarrow[+\overleftarrow{H}^{\ominus}]{-H^{\oplus}} AH^{\oplus}$$

A well-known example is acetylaminopyrenetrisulfonate, which loses a proton from the amino group to form the tetra-anion in the excited state (Weller, 1958). Absorption and fluorescence spectra of the neutral molecule, recorded in slightly acid solution, and of the tetra-anion, recorded in alkaline solution, are reproduced in Fig. 1.26. Many examples

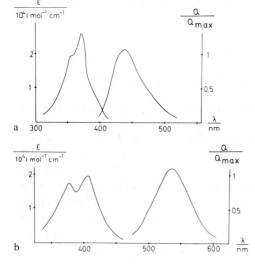

Fig. 1.26. Absorption and fluorescence spectra of acetylamino-pyrene-trisulfonate, a neutral solution, b alkaline solution (1 N NaOH)

of this type can be found among the numerous known fluorescence indicators which show a change in fluorescence wavelength and/or intensity with a change of pH of the solution. While the first example of a dye laser utilizing such a protolytic reaction in the excited state (SCHÄFER, 1968) was the above-mentioned acetylaminopyrenetrisulfonate pumped by a flashlamp, the first example used in a nitrogen-laser pumped dye laser and showing an especially large shift of the emission wavelength with pH was given by SHANK et al. (1970a, b) and DIENES et al. (1970). They used two coumarins which can form a protonated excited form that fluoresces at longer wavelengths. The chemistry of these dyes is discussed in detail in Chapter 4. By judicious adjustment of the acidity of an ethanol/water mixture, neutral and protonated forms of these dyes could be made to lase. The tuning extended over 176 nm (from $391-567$ nm). The term "exciplex", which these authors use for the excited protonated form, should be reserved for cases where an excited complex is formed with another solute molecule, not merely with a proton.

Donor-acceptor charge-transfer complex formation between a dye and a solvent molecule can occur in the ground state as well as in the excited state. The latter case is especially interesting, since the ground state of the complex can have a repulsion energy of a few kcal, while in the excited state the enthalpy of formation is comparable to that of a weak to moderately strong chemical bond [KNIBBE et al., 1968]. Thus the complex is stable in the excited state only and cannot be detected in the absorption spectrum. The rate of complex formation is limited by the diffusion of the two constituents. It is thus proportional to the product of the concentrations and strongly depends on the viscosity of the solution.

$$A^* + D \rightleftharpoons (A^\ominus D^\oplus)^*$$

$$\uparrow h\nu_{abs} \qquad \downarrow h\nu_{fl}$$

$$A + D \rightleftharpoons (A^\ominus D^\oplus) \longrightarrow A^\ominus + D^\oplus \longrightarrow \text{reaction products.}$$

If the excited complex $(A^\ominus D^\oplus)^*$ is nonfluorescing, the addition of D to A results merely in quenching of the fluorescence of A. This is generally believed to be a possible mechanism of the so-called dynamic quenching of fluorescence. On the other hand, if the excited complex is fluorescent, a new fluorescence band appears, while the original fluorescence disappears with increasing concentration of D, which might be one of the constituents of a solvent mixture. An example of this behavior is the complex formed from dimethylaniline and anthracene in the excited state (KNIBBE et al., 1968). While this principle has not yet been utilized in laser-pumped lasers, a methanol solution of a pyrylium dye in a flashlamp-pumped laser showed a significant lowering of the threshold and displacement of the laser wavelength by 10 nm to the red on addition of a small quantity of dimethylaniline, evidently as a result of the formation of a charge-transfer complex in the excited state.

Still another type of solvent influence on the dye can occur in such dyes as rhodanile blue

which can exist in the two conformations shown in Fig. 1.27. One would expect both conformations to show practically the same absorption spectrum, while the fluorescence spectrum of both conformations should

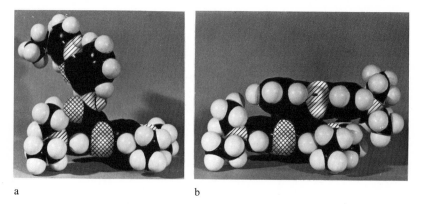

a   b

Fig. 1.27. Models of two possible conformations of the dye rhodanile-blue, a in hexafluoropropanol-(2), b in methanol

display a marked difference. For the conformation in Fig. 1.27a, both moieties of the molecule are practically decoupled and fluoresce independently, whereas the conformation of Fig. 1.27b should show a strong energy transfer from the rhodamine moiety to the nile-blue moiety. This is evidenced by the spectra of this dye in hexafluoroisopropanol and methanol shown in Fig. 1.28.

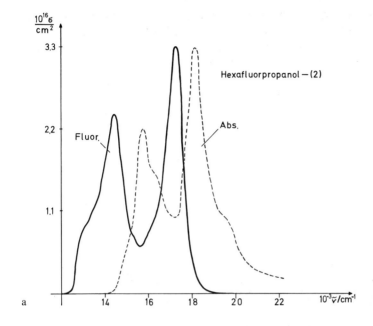

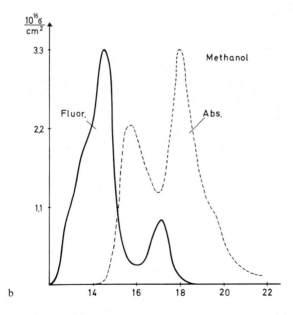

Fig. 1.28. Absorption and fluorescence spectra of the dye rhodanile-blue a in hexa-fluoropropanol-(2), b in methanol

Assuming complete absorption of the pump radiation for a specific dye (and at constant temperature), the energy conversion efficiency is strongly dependent on the solvent. This is evidenced by the results of SOROKIN et al. (1967) given in the following table of conversion efficiencies of an end-pumped DTTC laser with the concentration adjusted to absorb 70% of the ruby-laser pump radiation.

Solvent	Conversion Efficiency, %
Methyl alcohol	9.0
Acetone	7.0
Ethyl alcohol	6.5
1-propanol	11.5
Ethylene glycol	14.0
DMF	15.5
Glycerin	15.5
Butyl alcohol	9.0
DMSO	25.0

The wavelength coverage of laser-pumped dye lasers is determined on the short-wavelength side by the shortest available pump-laser wavelength of sufficiently high peak power; on the long wavelength side it is given by the stability of the dyes used. The shortest wavelength of a dye laser reported to date is 336–360 nm for p-terphenyl in cyclohexane or ethanol pumped by a nitrogen laser (ABAKUMOV et al., 1969a, b). The near-infrared range is well covered by 19 different cyanines and mixtures thereof in different solvents, reported by MIYAZOE and MAEDA (1968, 1970). These authors obtained laser emission over a continuous range from 710 – 1000 nm, at powers higher than 1 MW. Pumping a solution of a polymethine dye, whose formula was not disclosed, by a neodymium glass laser, VARGA et al. (1969) obtained a dye-laser emission at 1176 nm, the longest wavelength reported to date. At any wavelength between these extremes there are several dyes available that are suitable for laser-pumped dye lasers, and more can easily be found by slight chemical variations of the molecule as described in Chapter 4. A list of known laser dyes can be found at the end of Chapter 4.

1.5. Flashlamp-Pumped Dye Lasers

1.5.1. Triplet Influence

In the case of flashlamp-pumped lasers, triplet effects become important because of the long risetime or duration of the pump light pulse. Figure 1.29 shows the time dependence of the population density n_1 in

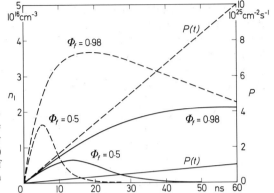

Fig. 1.29. Analog computer solutions of the population of the first excited singlet state n_1, for different pump powers $P(t)$ and quantum efficiencies of fluorescence, ϕ_f. (From SCHMIDT and SCHÄFER, 1967)

the excited singlet state (SCHMIDT and SCHÄFER, 1967). The solid curve applies to a slowly rising, and the broken curve to a fast rising pump light pulse. Both are computed for two different values of the quantum yield of fluorescence ϕ_f. It is assumed that there is no direct radiationless internal conversion between the first excited singlet and the ground state, so that a fraction $\varrho = 1 - \phi_f$ of excited molecules in the singlet state make an intersystem crossing to the triplet state. It is further assumed that the triplet state has a very long lifetime compared to the singlet state. It is seen that there is a maximum in the excited singlet state population density n_1, which is reached while the pump light intensity is still rising. This maximum is higher for a faster rise of the pump light and also for a higher quantum yield of fluorescence. Despite the continuing rise of the pump light intensity, n_1 falls to a low value after passing the maximum since the ground state becomes depleted and virtually all of the molecules accumulate in the triplet state. It should be pointed out that this behavior was computed for the spontaneous fluorescence of an optically thin layer of dye solution whose intensity is proportional to n_1. Obviously, this type of behavior would remain practically unaltered when stimulated emission is taken into account. Thus a dye laser may be pumped above threshold by a fast-rising light source. On the other hand, the necessary population density in the excited singlet state may not be achieved with a slowly rising light source, even if the asymptotic pump level is the same for both situations. In any case, the laser emission would be extinguished after some time. Even more important than this depletion of the ground state, however, are additional losses due to molecules accumulated in the triplet state. These give rise to triplet–triplet absorption spectra, as described in Section 1.3, which very often extend into the region of fluorescence emission. Therefore they can lead to losses that often are higher than the gain of the laser and eventually prevent laser emission.

Experimental evidence for the premature stopping of laser action due to triplet losses can be found in almost any flashlamp-pumped dye laser. One especially clearcut example is given in Fig. 1.30, which shows the pump and dye laser pulse of a small annular air flashlamp and a solution of the dye acetylaminopyrene-trisulfonate (Schäfer, 1970). While in oscillogram a) the solution is neutral, it is slightly alkaline in b). In both cases the absorption and hence pump rate and thermally induced

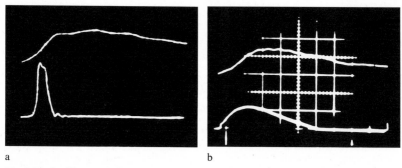

a   b

Fig. 1.30. Oscillograms of pumping-light pulse (upper beam) and dye-laser pulse (lower beam) for a 10^{-3} M aqueous solution of acetylaminopyrene-trisulfonate. Sweep speed 100 nsec/cm, a neutral solution, b alkaline solution (1 N NaOH). (From Schäfer, 1970)

schlieren are practically identical. Yet, in the neutral solution the laser pulse lasts only 250 nsec and is terminated before the pump pulse has reached its maximum, while in the alkaline solution the anion of the dye is the active species, as explained before, and there the pulse lasts 1.4 μsec.

Several authors have treated the kinetics of dye-laser emission by a set of coupled rate equations including terms to account for triplet–triplet absorption losses. Since the triplet losses are time-dependent, they affect the efficiency as well as the emission wavelength of the dye laser. Figure 1.31 gives the dye laser efficiency as a function of the excitation pulse width (Sorokin et al., 1968). The parameter a is the ratio of the total number of photons in the excitation pulse to the number of photons at threshold. The figure applies to the case $\phi_f = 0.88$ and $\sigma_a = 10\sigma_T$ and $\tau_T = \infty$, where σ_T is the cross-section for the triplet–triplet absorption band. Figure 1.32 is a plot of gain vs. wavelength calculated for a rhodamine B laser with time (in nsec) after initiation of the flashlamp pulse as a parameter (Weber and Bass, 1969). The results of such calculations apply only to the specific light pulse form considered. It is therefore worthwhile to have an expression connecting population densities

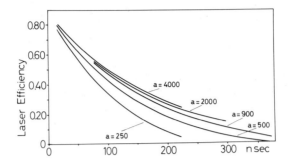

Fig. 1.31. Dye-laser efficiency vs. half-width of pumping pulse at constant pulse energy (parameter a). (From SOROKIN et al., 1968)

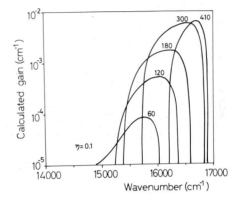

Fig. 1.32. Gain vs. wavelength calculated for a rhodamine B laser with time (in nsec) after initiation of the pump pulse as a parameter. (From WEBER and BASS, 1969)

in the ground, lowest excited singlet, and triplet states with the laser wavelength and cavity parameters.

The following simple modification of the oscillation condition of Subsection 1.4.1 takes into account triplet–triplet absorption losses. characterized by a cross-section σ_T. Let $\alpha = n_0/n$ be the normalized ground-state population density, and $\beta = n_T/n$ the triplet-state population density. Then the oscillation condition, its derivative with respect to wave number, and the balance of population densities give the following three equations for α, β, γ (the prime denotes differentiation with respect to wave number):

$$-\sigma_a\alpha - \sigma_T\beta + \sigma_f\gamma = S/n\,, \tag{1.12a}$$

$$-\sigma_a'\alpha - \sigma_T'\beta + \sigma_f'\gamma = 0\,, \tag{1.12b}$$

$$\alpha + \beta + \gamma = 1\,. \tag{1.12c}$$

From these equations and the observed laser wavelength, the population densities in the ground, excited singlet, and triplet states can be obtained. As an example of this procedure the case of rhodamine 6 G in a cw laser will be discussed in detail in Chapter 2. The ratio β/γ, obtained from the above relations, and the time t_0 to reach threshold can be used for an estimate of k_{ST}, if one assumes $t_0 \ll \tau_T$ and a linearly rising pump light source. Then we have $d\gamma/dt = $ constant and $d\beta/dt = k_{ST} \cdot \beta$, which yields after integration $\beta/\gamma = \frac{1}{2}k_{ST} \cdot t_0$. On the other hand, the observed t_0 can give an upper limit for σ_T at the laser wavelength for known k_{ST}, since laser emission can only be achieved for $\gamma\sigma_f > \beta\sigma_T$, yielding $\sigma_T > 2\sigma_f/t_0 k_{ST}$ (Sorokin et al., 1968).

For pulses that are long compared with τ_T or for continuous operation, a steady state must be reached. Then the triplet production rate $k_{ST}\gamma$ equals the deactivation rate β/τ_T, so that the ratio $\beta/\gamma = k_{ST}\tau_T$. Setting this value equal to that obtained above from the observed laser wavelength, an estimate of τ_T can be obtained.

To obviate the need for a rapidly rising pump light intensity in dye lasers and to achieve cw emission, one must reduce the triplet population density to a sufficiently low level by reducing τ_T. One way of achieving this is by adding to the dye solution suitable molecules that enhance the intersystem-crossing rate $k_{TG} = 1/\tau_T$ from the triplet to the ground state by the effects described in Section 1.3. The improvement in dye-laser performance possible with this method was shown by Keller (1970), who solved the complete rate equations of Sorokin et al. (1968) after adding a term n_T/τ_T for the deactivation of the triplet state to the equation for the triplet-state population density.

A molecule that has long been known to quench triplets is O_2. At the same time, however, it also increases k_{ST} and thus also quenches the fluorescence of a dye. The relative importance of triplet and fluorescence quenching then determines whether or not oxygen will improve dye-laser performance. This is shown in Fig. 1.33, which gives the output from a long-pulse rhodamine 6 G laser as a function of the partial pressure of oxygen in the atmosphere in contact with the dye solution in the reservoir (Schäfer and Ringwelski, 1973). While the output first rises steeply with increasing oxygen content, it levels off after about 20% oxygen content, showing that the positive effect of the reduction in triplet lifetime is offset by the detrimental effect of fluorescence quenching. This can be understood quantitatively, if one assumes that the laser output is proportional to the ratio γ/β. In the presence of an oxygen concentration $[O_2]$ the triplet production rate is $(k_{ST} + k_{QS}[O_2])n_1$ and is set equal to the deactivation rate $(k_{TG} + k_{QT}[O_2])n_T$, where k_{QS} and k_{QT} are the quenching constants for the singlet and triplet states, respectively. This yields $\gamma/\beta = (k_{TG} + k_{QT}[O_2])/(k_{ST} + k_{QS}[O_2])$. Using $k_{ST} = 2 \cdot 10^7 \, \mathrm{sec}^{-1}$, $k_{QS} = 3 \cdot 10^{10} \, \mathrm{sec}^{-1} \cdot 1 \cdot \mathrm{mole}^{-1}$, and $k_{QT} = 3.3 \cdot 10^9 \, \mathrm{sec}^{-1} \cdot 1 \cdot \mathrm{mole}^{-1}$, we find good agreement with the experimental results for $k_{TG} = 5 \cdot 10^5 \, \mathrm{sec}^{-1}$. The triplet lifetime of rhodamine 6 G in carefully deaerated methanol

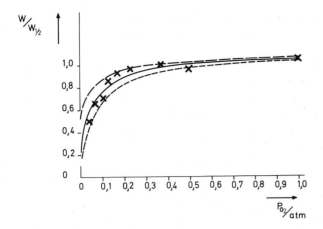

Fig. 1.33. Normalized laser output $W/W_{1/2}$ vs. oxygen partial pressure P_2 in the gas mixture over a solution of rhodamine 6 G. (Crosses: measurements; solid line: theoretical result for $k_{TG} = 5 \cdot 10^5$ sec^{-1}; upper dashed curve: same, but for $k_{TG} = 10^6$ sec^{-1}; lower dashed curve: same for $k_{TG} = 10^4$ sec^{-1}). (From Schäfer and Ringwelski, 1973)

solutions at room temperature is thus $\tau_T = 2\,\mu$sec, while the effective triplet lifetime with 20% oxygen content in the atmosphere above the dye is $\tau_{T_{eff}} = 1/(k_{TG} + k_{QT}[O_2])$ or 140 nsec. Marling et al. [1970a] measured the effect of molecular oxygen at different partial pressures on the dye laser emission of a number of dyes. They found several dyes, where the fluorescence quenching of oxygen was much stronger than its triplet quenching effect, e.g. the dye brilliant sulphaflavine, whose laser emission was extinguished when the argon atmosphere in the reservoir was replaced by one atmosphere of oxygen.

It is thus seen to be more appropriate not to use a paramagnetic gas like oxygen, which enhances both k_{ST} and k_{TG}, but rather to apply energy transfer from the dye triplet to some additive molecule with a lower lying triplet that can act as acceptor molecule. Triplet–triplet energy transfer was shown by Terenin and Ermolaev [1956] to occur effectively with unsaturated hydrocarbons. This scheme was used by Pappalardo et al. [1970b], who used cyclooctatetraene as acceptor molecule and obtained a dye-laser output pulse from rhodamine 6 G as long as 500 µsec, demonstrating that the triplet lifetime had been reduced by the quenching action of cyclooctatetraene below the steady-state value necessary for cw-laser operation. This compound is still the most effective triplet quencher known for rhodamine 6 G, although a number of others have been tested and several quite effective ones found [Marling, 1970b], e.g. N-aminohomopiperidine, 1,3-cyclooctadiene, and the nitrite ion, NO_2^-.

It must be stressed that this energy transfer occurs in such a way that energy as well as spin is exchanged between dye and acceptor molecule, leaving the acceptor in the triplet state, according to $^3D + {}^1A \rightarrow {}^1D + {}^3A$. MARLING (1970b) has pointed out that there is also a possibility of quenching a dye triplet by collision with a quencher molecule in the triplet state through the process of triplet–triplet annihilation, which leaves the dye in the excited singlet state and the quencher molecule in the ground state: $^3D + {}^3A \rightarrow {}^1D^* + {}^1A$ with subsequent fluorescence emission or radiationless deactivation of the excited singlet state. In this case, the quencher molecules must first be excited to the first excited singlet state and pass by intersystem crossing to the triplet state, before they can become effective. The first excited singlet state of quencher molecules must always lie higher than that of the dye, so that no pump radiation useful for pumping the dye is absorbed and no fluorescence quenching energy transfer can occur from the first excited singlet state of the dye to that of the quencher molecule. This means that, for efficient quencher molecule triplet production, the pump light source must have a high proportion of ultraviolet light output. Usually this means at the same time a higher photodegradation rate for the dye, which also absorbs part of the ultraviolet pump light. Because of these disadvantages and its generally lower effectiveness, this indirect method of triplet quenching seems less attractive. The distinction between direct and indirect triplet quenching can, however, clarify the results of other experimental investigations [STROME and TUCCIO, 1971; SMOLSKAYA and RUBINOV, 1971].

1.5.2. Practical Pumping Arrangements

Flashlamp-pumped dye laser heads consist in principle of the dye cuvette, the flashlamp and a pump light reflector or diffuser. The latter serves to concentrate the pump light emitted from the extended, un-collimated, broadband source, the flashlamp, onto the absorbing dye solution in the cuvette. The reflector can be of the imaging type, e.g. an elliptical cylinder whose focal lines determine the position of linear lamp and cuvette, or it can be of the close-coupling type, which is especially advisable where there are several flashlamps surrounding the cuvette. Instead of a specular reflector, a diffusely reflecting layer of MgO or $BaSO_4$ behind a glass tube surrounding flashlamp and cuvette is often used. The design of the pump cavity is thus similar to that of solid-state lasers, except that for dye lasers it is even more important to prevent nonuniform heating of the solution in order to avoid thermally induced schlieren. Furthermore, it is advisable to use some means of filtering out photochemically active wavelengths which might decompose the dye molecules. It is often sufficient to use an absorbing glass tubing for the cuvette, otherwise a double-walled cuvette or a filter solution surrounding the flashlamp can be used.

For maximum utilization of the pumplight output the length of the cuvette will generally be about the same as that of the flashlamp. This in turn makes a flow system almost mandatory, because in a long cuvette even small thermal gradients can severely degrade the resonator characteristics. The dye flow in the cuvette may be longitudinal or transverse. In either case the flow should be high enough to be in the turbulent regime. This rapidly mixes the liquid and hence reduces thermal gradients due to nonuniform pump-light absorption in the cuvette. Both flow orientations are used in high-repetition-rate dye lasers (BOITEUX and DE WITTE, 1970; SCHMIDT, 1970). If the cuvettes are not of all-glass construction, windows are generally pressed against the cuvette end and sealed by O rings. These O rings, and the hoses connecting the cuvette with the circulating pump and the reservoir, can be of silicone rubber for use with methanol or ethanol solvents. Commercial O rings, and even clear plastic hoses, usually give off absorbing or quenching filler material and hence should not be used. For other solvents, like dimethylformamide or dimethylsulfoxide, Teflon coatings are required on O rings and hoses. A similar choice is required with the circulating pump and the reservoir for the dye solution. Magnetically coupled centrifugal or toothed wheel pumps made of Teflon or stainless steel are well suited for this application, whereas membrane pumps give less reproducible results because of the pulsation in the flow rate. The metal end pieces carrying the nipples for the inflow and outflow of the dye solution, situated between the cuvette made of glass tubing and the windows, are also best made of stainless steel and should contain as little unpumped length of dye solution as possible so as to reduce the reabsorption of the dye-laser beam. As an example, the cross-sections of a small dye-laser head for 100 Hz pulse-repetition frequency and several watts average output power are given in Fig. 1.34. A great variety of flashlamps have been used in dye lasers. The simplest possibility is the use of commercial xenon flashlamps as in the laser head of Fig. 1.35. The risetime of linear and helical flashlamps can be reduced and the output power increased by the introduction of a spark gap in series with the lamp; this allows the lamp to be operated at a voltage that is much higher than the self-firing voltage. The excess voltage applied to the lamp ensures a rapid build-up of the plasma and a much higher peak power in the lamp. In this arrangement with a linear, 8-cm-long xenon lamp of 5 mm bore diameter and a low-inductance 0.3 μF/20 kV capacitor, a risetime of 300 nsec can be obtained. By comparison, a helical flashlamp of 8 cm helix length and 13 mm inner

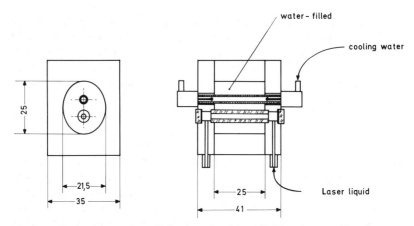

Fig. 1.34. Cross-sections of small dye-laser head for 100 Hz-pulse repetition frequency (dimensions in mm). (From SCHMIDT, 1970)

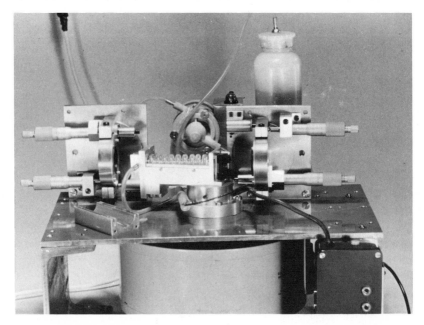

Fig. 1.35. Photograph of a dye laser with helical flashlamp

diameter of the helix gives a risetime of about 0.5 μsec. For reasons connected with the accumulation of molecules in the triplet state, discussed in the last section, much effort has gone into the development of high-power lamps of short risetime. Very fast risetimes (70 nsec) can be achieved by the use of small capillary air sparks fed by an energy storage capacitor in the form of a flat-plate transmission line (AUSSENEGG and SCHUBERT, 1969). Another type with a very fast risetime is a low-inductance coaxial lamp in which the cylindrical plasma sheet surrounds the cuvette. This type was first developed for flash photolysis work (CLAESSON and LINDQUIST, 1958). Later it was used successfully for dye-laser pumping (SOROKIN and LANKARD, 1967; SCHMIDT and SCHÄFER, 1967). An improved version of this lamp was developed by FURUMOTO and CECCON (1969). With this lamp too a spark gap is used in series with the lamp, so that a voltage much higher than the breakdown voltage of the lamp can be used. At the same time the pressure can be adjusted so that the plasma fills the lamp uniformly. By comparison the original design shows constricted spark channels which move from shot to shot. The improved version offers nearly uniform illumination of the cuvette and the pulse height is reproducible from shot to shot. Risetimes of 150 nsec using a 0.05 μF capacitor were achieved with this lamp. For smaller capacitors even shorter risetimes can be obtained.

This configuration is also amenable to up-scaling, and this has been done by Russian workers (ALEKSEEV et al., 1972; BALTAKOV et al., 1973) who obtained dye-laser pulses of up to 150 joule output energy. Figure 1.36 shows a cross-section through a dye laser using such a lamp.

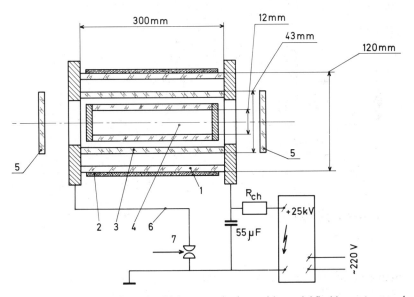

Fig. 1.36. Cross-section through a high energy dye laser with coaxial flashlamp. *1* external tube, *2* silicon dioxide coating, *3* internal tube, *4* dye cuvette, *5* mirrors, *6* external current lead, *7* triggered spark gap, R_{ch} = charging resistor. (From ALEKSEEV et al., 1972)

1.5.3. Time Behavior and Spectra

The time behavior of flashlamp-pumped dye lasers is more complex than
that of laser-pumped dye lasers because of time-dependent triplet losses
and thermally or acoustically induced gradients of the refractive index.
Several authors have derived solutions of the rate equations for flashlamp-pumped dye lasers under various approximations (Bass et al.,
1968; Weber and Bass, 1969; Sorokin et al., 1968). In view of the many
quantitative uncertainties, a calculation of the time behavior of flashlamp-pumped dye lasers is of rather limited value. Instead, experimental
results are given here.

For small coaxial lamps having a fast risetime and short pulse widths,
the time behavior is similar to that of laser-pumped dye lasers. Thus,
Furumoto and Ceccon (1970) obtained a pulse of 40 kW peak power
and 100 ns duration in the ultraviolet from a solution of p-terphenyl in
DMF using a lamp of 50 ns risetime and 20 J energy capacity. With
similar lamps Hirth et al. (1972) and Maeda and Miyazoe (1972) were
able to obtain laser emission from many cyanines in the visible and near-
infrared.

Lasers equipped with commercial xenon flashlamps have slower risetimes. Consequently the number of dyes that will lase in these devices
is restricted and triplet quenchers must be used if long pulse emission is
wanted. Nevertheless, high average and peak powers and relatively high
conversion efficiencies can be obtained in this way. Figure 1.37 shows a
plot of dye-laser peak power versus flashlamp energy for the small laser
head shown in Fig. 1.34 (Schmidt, 1970). An optimized version allowed
operation at 100 Hz pulse repetition frequency and an average dye-laser
output power of 3.5 W (Schmidt and Wittekindt, 1972). With linear
flashlamps in an elliptical pump cavity powers of up to 1 MW were
obtained (Bradley, 1970).

With a lumped constant transmission line in place of a single
capacitor, a long, flat-top pump-light pulse can be formed. Figure 1.38
is an oscillogram of such a pump-light pulse from a helical flashlamp and
the dye-laser pulse excited by it (Ringwelski and Schäfer, 1970). A
400 μsec long dye-laser pulse is obtained from an air-saturated methanol
solution of rhodamine 6 G. Here, proper filtering of the pumplight
through a copper sulfate solution was required to reduce unnecessary
heating. Also, an optimal concentration of the dye was chosen, so that
heating due to pumplight absorption was reasonably uniform throughout
the cuvette volume. With the same pumping arrangement Snavely and
Schäfer (1969) had obtained 140 μsec long pulses from rhodamine 6 G
and rhodamine B solutions; their duration clearly indicated that a steady
state of the triplet population had been reached, provided the solution

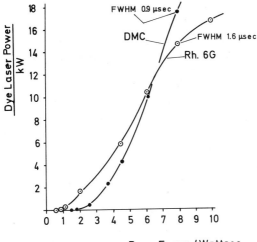

Fig. 1.37. Dye laser peak output power vs. flashlamp, pulse energy at 100 Hz pulse repetition frequency for rhodamine 6 G (Rh. 6 G) and 7-diethylamino-4-methylcoumarin (DMC). (From SCHMIDT, 1970)

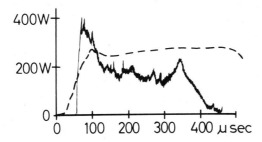

Fig. 1.38. Oscillogram of pumplight pulse (broken-line) of a dye laser using a helical flashlamp and a lumped parameter transmission line and dye-laser pulse (solid line) from an air-saturated rhodamine 6 G solution

was saturated with oxygen or air. No laser emission was obtained, even at twice the original threshold, if nitrogen was bubbled through the solution for a time sufficient to purge the oxygen. PAPPALARDO et al. (1970b) using a laser with a 600-μsec pump pulse obtained a 500-μsec dye laser pulse from a $5 \cdot 10^{-5}$ molar rhodamine 6 G solution containing $5 \cdot 10^{-3}$ mole/l of cyclooctatetraene as triplet quencher. Such results with long pulses first proved that triplet absorption cannot prevent cw dye-laser emission if the necessary triplet quenchers are added. In fact, it was concluded from these experiments that an absorbed pump-light power of less than $4\,kW/cm^3$ for a $5 \cdot 10^{-5}$ molar air-saturated methanol solution of rhodamine 6 G should be sufficient to reach threshold with steady-state triplet population. The thermal problem, on the other hand, remains a serious one. To alleviate this problem one might employ a solvent with higher specific heat and use it at a temperature where the variation of the refractive index with temperature is at a minimum. In this respect water near freezing point or, even better, heavy water at $6°$ C would be an ideal solvent.

Extremely high dye-laser pulse energies have been obtained with large coaxial lamps. The laser shown in Fig. 1.36 gives an output energy of 32 J with an alcoholic rhodamine 6 G solution at 17.3 kJ electrical energy input, i.e. 0.2% efficiency, and a specific output energy of $1\,J/cm^3$. The output power was reported as 10 MW, so that the half-width of the pulse must be about 3.2 µsec. The lamp was filled with xenon at 1 Torr pressure in this case. The laser reported by BALTAKOV et al. (1973) generated 110 J in 20 µsec, and 5.5 MW peak power.

The spectral range of flashlamp-pumped dye lasers at present extends from 340 nm in p-terphenyl to 850 nm in DTTC (MAEDA and MIYAZOE, 1972). Time-resolved spectra of flashlamp-pumped dye lasers show even greater variety than those of laser-pumped dye lasers (FERRAR, 1969a). Some dyes show almost no wavelength sweep during an emission of 300 nsec, while others have either monotonic or reversing sweeps. In these experiments triplet–triplet absorption and thermal effects due to nonuniform illumination of the cuvette may have been of importance. If the triplet–triplet spectrum is known, the sweep can be predicted. Thus in rhodamine 6 G a sweep towards shorter wavelengths should be observed in a uniformly illuminated laser cuvette (SNAVELY, 1969).

In addition to these fast sweeps there is a long-term drift in wavelength associated with increasing loss due to products of photochemical decomposition of the dye. Thus, in a 100 Hz flashlamp-pumped rhodamine 6 G laser, the laser wavelength was observed to drift from 570–600 nm in a 10 min period (SCHMIDT, 1970). The absorption spectrum of the solution taken after this experiment gave clear evidence of a photodecomposition product with absorption increasing towards shorter wavelengths.

1.6. Wavelength-Selective Resonators for Dye Lasers

A coarse selection of the dye-laser emission wavelength is possible by judicious choice of the dye, the solvent, and the resonator Q, as described in Subsection 1.4.1. Fine tuning and simultaneous attainment of small linewidths can only be achieved by using a wavelength-selective resonator.

Up to now the following four classes of wavelength-selective resonators seem to have been employed:

1) resonators including devices for spatial wavelength separation,

2) resonators including devices for interferometric wavelength discrimination,

3) resonators including devices with rotational dispersion,

4) resonators with wavelength-selective distributed feedback.

The various implementations of these classes of wavelength-selective resonators and their relative merits will be discussed in the above order.

The first wavelength-selective resonator was constructed by SOFFER and McFARLAND (1967). They replaced one of the broad-band dielectric mirrors by a plane optical grating in Littrow mounting. This arrangement is shown in Fig. 1.39 together with a diagram of laser output vs. wave-

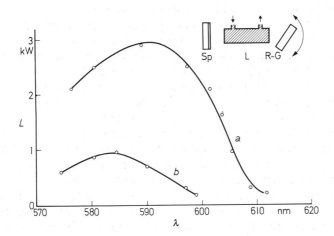

Fig. 1.39. Tuning of dye-laser wavelength with an optical grating. Output powers vs. wavelength for a 10^{-4} M solution of rhodamine 6 G in methanol, grating 610 lines/mm, a first order, b second order. (Insert: optical arrangement; Sp = 98 % mirror, L = laser cuvette, R-G = optical grating in rotatable mount)

length obtained with a grating of 610 lines per mm in the first and second order for a rhodamine 6 G solution. Consider the grating equation $m\lambda = d(\sin\alpha + \sin\beta)$ which for autocollimation ($\alpha = \beta$) reduces to $m\lambda = 2d\sin\alpha$. Here m is the order, λ is the wavelength, β is the angle of incidence, α is the angle of diffraction from the normal to the grating, and d is the grating constant. Then the angular dispersion is $d\alpha/d\lambda = m/2d\cos\alpha$.

If the dye laser has a beam divergence angle $\Delta\alpha$, the passive spectral width of this arrangement would be

$$\Delta\lambda_\alpha = \frac{2d\cos\alpha}{m}\Delta\alpha\,. \tag{1.13}$$

On the other hand, if the laser has diffraction-limited beam divergence, $\Delta\alpha = 1.22\lambda/D$, where D is the inner diameter of the cuvette, then the passive spectral width is

$$\Delta\lambda_D = 2.44\,d\lambda\cos\alpha/mD\,. \tag{1.14}$$

For a typical case, $\Delta\alpha = 5$ mrad, $D = 2.5$ mm, $d = (1/1200)$ mm, $m = 1$, and $\lambda = 600$ nm, one obtains $\Delta\lambda_\alpha = 7.8$ nm, and $\Delta\lambda_D = 0.37$ nm. If beam-expanding optics are used within the cavity to reduce the beam divergence by a factor of ten, this would give $\Delta\lambda_{\alpha r} = 0.78$ nm and $\Delta\lambda_{Dr} = 0.037$ nm. The active spectral width at threshold is smaller than the passive width depending on the available gain. Passive bandwidths are quoted here since these give an upper limit for the spectral bandwidth. The experimental value of the spectral width for a rhodamine 6 G laser with 5 mrad beam divergence and a 2160 lines/mm grating was 0.06 nm, as compared with a passive width of 4.6 nm (Soffer and McFarland, 1967).

For another laser with a 600 lines/mm grating the spectral width was 2 nm in the first and 0.4 nm in the second order, as compared with the passive width of 16 nm and 8 nm, respectively (Marth, 1967). Thus the active bandwidth is smaller by a large factor (between 8 and 80 in these cases) than the passive bandwidth. The peak power is reduced by a factor of only two to five if the grating is blazed for this wavelength. In a laser with a Brewster angle cuvette and resultant polarized emission, this ratio can be even more favourable for the polarization which gives higher grating efficiency.

While high-quality gratings can have efficiencies of up to 95% at the blaze wavelength, most gratings have lower efficiencies, 65% being a realistic value. Thus, the insertion loss due to the grating is substantial.

Another disadvantage of the grating is the reflecting metal film which may be damaged by high power and energy pulses. An improvement may be expected from holographically produced bleached transmission gratings (Kogelnik et al., 1970). This problem can also be circumvented by the use of a high-power beam-expanding telescope. The use of such a beam-expanding device is especially indicated in the case of nitrogen laser-pumped dye lasers, as shown in Fig. 1.20. An unexpanded beam of a fraction of one mm diameter would cover only a few lines on the grating and thus also seriously impair the spectral resolution. Another way to prevent the burning of a grating is by the use of an additional semitransparent mirror in front of the grating, as shown in Fig. 1.40 (BJORK-HOLM et al., 1971). This scheme not only reduces the power incident on the grating to a few percent of what it would have been without mirror, but also significantly reduces the laser threshold, since the mirror–grating combination acts as a high-reflectivity resonant reflector for the tuned wavelength. The authors report a reduction in threshold by a factor of two and in bandwidth by a factor of 3.3 over the use of a grating only.

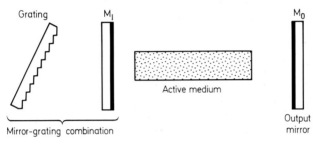

Fig. 1.40. Schematic diagram showing the use of a mirror-grating combination in a laser. M_1 is the intermediate mirror. (From BJORKHOLM et al., 1971)

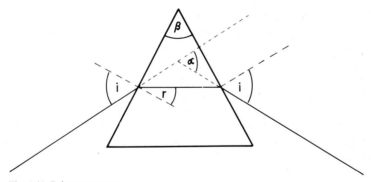

Fig. 1.41. Prism geometry

Alternatively, tuning and spectral narrowing may be achieved by one or more prisms in the laser cavity (YAMAGUCHI et al., 1968). The relatively small angular dispersion of a single prism is sufficient to isolate one of several sharp lines in gas lasers, for example, where this method has long been used. But it gives hardly any reduction in spectral bandwidth of a flashlamp-pumped dye laser, so that multiple-prism arrangements have to be used. With the notation of Fig. 1.41 one has $\alpha = 2i - \beta$ and $r = \frac{1}{2}\beta$ so that

$$\frac{d\mu}{d\alpha} = \frac{\cos\frac{1}{2}(\alpha + \beta)}{2\sin\frac{1}{2}\beta} \ . \tag{1.15}$$

Since it is better to work near the Brewster angle where $d\alpha/d\mu = 2$, the angular dispersion of a prism is

$$d\alpha/d\lambda = 2d\mu/d\lambda . \tag{1.16}$$

Using z prisms in autocollimation with a dye laser of beam divergence $\Delta\alpha$, the passive spectral width is

$$\Delta\lambda_\alpha = \frac{\Delta\alpha}{4zd\mu/d\lambda} . \tag{1.17}$$

If $\Delta\alpha = 1.22\,\lambda/D$, in the diffraction-limited case one has

$$\Delta\lambda_D = \frac{1.22\lambda/D}{4zd\mu/d\lambda} . \tag{1.18}$$

Consider 60°-prisms of Schott-glass SF 10 for which $\mu_D = 1.72802$ and $d\mu/d\lambda = 1.35 \times 10^{-4}\,\text{nm}^{-1}$. Then the following values are obtained for 1 or 6 prisms in autocollimation:

Values of $\Delta\lambda$ [nm] for a number of prisms

No. of prisms	$\Delta\lambda_\alpha$	$\Delta\lambda_{\alpha r}$	$\Delta\lambda_D$	$\Delta\lambda_{Dr}$
1	9.3	0.93	0.54	0.05
6	1.5	0.15	0.09	0.01
grating 1200 lines/mm	7.8	0.78	0.37	0.04

As above, it is assumed that $\Delta\alpha = 5$ mrad and $D = 2.5$ mm, with and without tenfold beam expanding optics.

A comparison of passive spectral widths, as given in the above table, shows that it should be better to use two or more prisms rather than one grating. In addition, the cumulative insertion loss of even 6 prisms near the Brewster angle is much smaller than that of one grating.

A five-prism arrangement has been reported by STROME and WEBB (1971) and a six-prism arrangement by SCHÄFER and MÜLLER (1971). Another method for obtaining increased angular dispersion uses a prism at angles of incidence of slightly less than 90 degrees. This results in a very high dispersion at only slightly reduced resolution, as in early spectrographs. The high reflection at the prism face that was detrimental in those spectrographs is an advantage here, since it serves as an out-coupling device, variable between 20% reflection at 80° incidence to 100% reflection at 90° (MYERS, 1971). In the prism there is a considerable increase in beam width and a concomitant reduction in beam divergence, which makes the additional insertion of a grating attractive, again particularly in nitrogen-laser-pumped dye lasers.

The prism method lends itself to use in ring lasers. SCHÄFER and MÜLLER [1971] incorporated six 60°-prisms made of the Schott-glass SF 10 as dispersive elements in a ring laser whose optical path is completed by two glass plates connected by a rubber bellows which is filled with an index-matching fluid so that a variable wedge is formed. The laser wavelength is selected by the setting of the refracting angle of the wedge (Fig. 1.42). A linewidth of less than 0.05 nm was obtained with

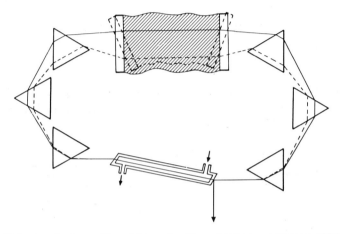

Fig. 1.42. 6-prism ring laser with variable wedge. (From SCHÄFER and MÜLLER, 1971)

40 J electrical input energy pumping a 10^{-4} molar rhodamine 6 G solution flowing in a 7.5-cm-long dye cuvette of 2.5 mm inner diameter, placed in the center of a 2″ helical flashlamp. The Fresnel reflection of one of the dye cuvette windows, intentionally set a few degrees off the Brewster angle, generated the output beam. It is noteworthy that the spectrum showed none of the satellite lines that are usually found in linear lasers using prisms or gratings for spectral narrowing. These satellite lines near the selected wavelength are probably associated with rays passing through inhomogeneities in the dye solution. Hence they form an angle with the optical axis which may be greater than the beam divergence of the main part of the beam. Insertion of an aperture into the cavity for reducing the beam divergence eliminates these lines. In a ring laser the cuvette evidently acts in this way, resulting in much improved discrimination against satellites. Another ring laser using four Abbé or Pellin-Broca prisms was described by MAROWSKY et al. [1972]. The four 90° constant-deviation prisms made of SF 10 glass are so arranged at the corners of a square that a closed path exists for the selected wave-

length. The wavelength tuning was achieved by simultaneous counter rotation of the four mechanically coupled prisms. The spectral range covered reached from 430 nm to beyond 700 nm and the spectral width was 0.8 nm at 600 nm and decreased towards shorter wavelengths (MAROWSKY, 1973a). This type of ring laser is shown in Fig. 1.43.

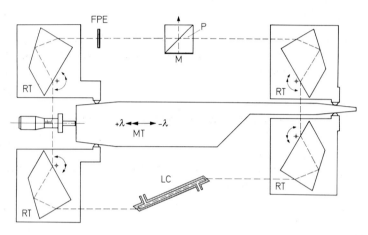

Fig. 143. Ring-laser. (LC laser cuvette, RT: rotating prism tables with Abbé prisms, axis and sense of rotation indicated, MT movable table, FPE: Fabry-Perot etalon, P beam-splitting output prism with high reflectivity mirror M). (From MAROWSKY et al., 1972)

A noteworthy advantage of these multiprism ring lasers is that they obviate the need for mirrors with broadband dielectric reflective coatings.

If only fixed wavelengths are needed, the simplest way of utilizing interference effects is to have narrowband reflective coatings on the resonator mirrors. Often supposedly broadband reflectors show a certain selectivity. At wavelengths where the reflectivity drops slightly, e.g. a quarter of one percent, holes are produced in the broadband spectrum of the dye laser at this wavelength; these holes have sometimes been misinterpreted as due to some property of the dye molecules. Resonant reflectors and reflectors of the Fox-Smith type are useful only in conjunction with suitable preselectors because of their narrow free spectral range.

In a method which is especially well suited for achieving a small spectral bandwidth, one or more Fabry-Perot etalons or interference filters are inserted into the cavity (BRADLEY et al., 1968). The wavelength λ of maximum transmission in kth order for a Fabry-Perot of thickness d, refractive index μ, and with an angle α between its normal and the optical axis, is given by $k\lambda = 2d/\mu \cos\alpha'$. Here α' is the refracted angle

$\mu \sin \alpha' = \sin \alpha$. Thus, for air ($\mu = 1$) the angular dispersion is

$$d\lambda/d\alpha = \lambda \tan \alpha. \qquad (1.19)$$

Hence the spectral bandwidth for beam divergence $\Delta \alpha$ is (independent of the finesse)

$$\Delta \lambda_\alpha = \lambda \Delta \alpha \tan \alpha . \qquad (1.20)$$

The wavelength shift $\Delta \lambda_s$ for turning the Fabry-Perot from a position normal to the optical axis ($\alpha = 0$, corresponding to a wavelength λ_0) through an angle α is

$$\Delta \lambda_s = (1 - \cos \alpha) \lambda_0 . \qquad (1.21)$$

The free spectral range $\Delta \lambda_F$ between adjacent orders is

$$\Delta \lambda_F = \lambda/k . \qquad (1.22)$$

The spectral width near λ_0 is determined by the reflection coefficient R of the Fabry-Perot mirrors,

$$\delta \lambda = \Delta \lambda_F/F , \qquad (1.23)$$

where

$$F = \frac{\pi \sqrt{R}}{1 - R} \qquad (1.24)$$

is the so-called finesse factor.

From these relations it is easy to determine the required properties of the laser and the Fabry-Perot for narrow band emission and wide band tunability. The attainable minimum bandwidth is determined by the minimum angle which avoids reflection from the first mirror of the Fabry-Perot back into the cuvette. If q is the ratio of the diameter of the cuvette to the distance between the Fabry-Perot and the nearest cuvette window, the minimum bandwidth is

$$\Delta \lambda_{\alpha \min} = \lambda \Delta \alpha \tan \tfrac{1}{2} q . \qquad (1.25)$$

Assume a prism preselector in the cavity which gives $\Delta \lambda_\alpha = 3$ nm for $\Delta \alpha = 5$ mrad at 600 nm. Then the optimum free spectral range of the Fabry-Perot should be $\Delta \lambda_F = \lambda/k = 3$ nm so that $k = 200$ and $d = 60$ μm. With a typical value of $q = 0.05$ this yields a beam-divergence-limited bandwidth of $\Delta \lambda_{\alpha \min} = 0.075$ nm. As the angle is increased to tune over the free spectral range, the bandwidth increases according to (1.20) to $\Delta \lambda_{\alpha \max} = 0.3$ nm. Thus, a large tuning range is possible only at the expense of a relatively large increase in bandwidth. In addition to an increasing bandwidth, the use of Fabry-Perot etalons at high angles also introduces serious walk-off losses, which become more serious the larger the ratio of etalon thickness to beam diameter and the higher the angle. In order to realize a specified narrow bandwidth, one would have to reduce the tuning range and/or the beam divergence of the laser. Wavelength selection and simultaneous spectral narrowing were achieved in this way with laser-pumped and flashlamp-pumped lasers. An emission bandwidth of less than 50 pm was first achieved in practice (Bowman et al., 1969a), later less than 1 pm (Schmidt, 1970). Hänsch (1972) reported that the insertion of a Fabry-Perot etalon into the dye laser cavity shown in Fig. 1.20c reduced the bandwidth from 3 pm, obtained with the grating only, to 0.4 pm. At the same time the output drops from 20 kW to about 3 kW, thought to be primarily due to high losses in the available broadband coatings. Probably at least some of these losses are due to other sources, like the walk-off in the etalon. This is substantiated by the small reduction of output powers after the successive insertion of an interference filter (low-order Fabry-Perot etalon) and two Fabry-Perot etalons used near the minimum useful angle. In his rhodamine 6 G laser at 600 nm wavelength and 20 J pump energy Marowsky (1973b) obtained 2.4 kW at 50 pm bandwidth with only an interference filter in the cavity. The insertion of the first Fabry-Perot decreased the peak power to 2.2 kW and the bandwidth to 7 pm. The insertion of the second etalon decreased the peak power to 2.0 kW and the bandwidth to 0.1 pm.

As Fabry-Perot etalons one usually uses either plane-parallel quartz plates coated with dielectric multilayer broadband reflective coatings, or optically contacted air Fabry-Perots (Bates et al., 1966).

Several methods of wavelength selection make use of the rotation of polarization. One method utilizes birefringent filters in the cavity (Soep, 1970). A simple arrangement consists of a quartz plate cut parallel to the optic axis, which has a retardation of several half-wavelengths at the center of the tuning range, and a set of Brewster plate polarizers as shown in Fig. 1.44. In this case there are transmission maxima for

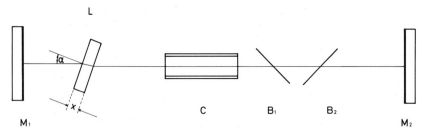

Fig. 1.44. Dye laser with birefringent filter. ($M_{1,2}$ mirrors, C dye cuvette, L quartz plate of thickness x, $B_{1,2}$ Brewster angle polarizers). (From Soep, 1970)

retardations of multiple half-wavelengths,

$$k\lambda/2 = \Delta\mu x_0 \cos\alpha .\qquad (1.26)$$

Here k is the order number, $\Delta\mu$ the birefringence and x_0 the crystal thickness, both for normal incidence. Thus the wavelength spread for beam divergence $\Delta\alpha$ is

$$\Delta\lambda = -\lambda \tan\alpha\, \Delta\alpha = -\lambda\, \frac{\cos\alpha \sin\alpha}{\mu^2 - \sin^2\alpha}\, \Delta\alpha .\qquad (1.27)$$

This expression is very small near $\alpha = 0$.

Now, however, the bandwidth is not determined by the beam divergence of the laser, as in the methods discussed above, but rather by the transmission T of the birefringent filter.

$$T = \cos^2\left(\pi\, \Delta\mu d/\lambda\right) .\qquad (1.28)$$

If a reduction of 10% in transmission compared to maximum transmission brings the laser below threshold, one would expect an active bandwidth of

$$\Delta\lambda_\alpha = \tfrac{1}{6}\lambda^2/\Delta\mu d .\qquad (1.29)$$

This spectral width can be reduced further by the introduction of one or more additional quartz plates of greater thickness, as in Lyot or similar birefringent filters. Using KDP crystals of suitable orientation instead of the quartz plates, one can vary the transmission wavelength by applying a voltage (WALTHER and HALL, 1970). The actual tuning range in a flashlamp-pumped rhodamine 6 G laser with a quartz plate of 0.36 nm thickness and an angle of incidence between $35°$ and $50°$ was from 570–600 nm with a spectral bandwidth of 1 nm (SOEP, 1970). WALTHER and HALL (1970) obtained a spectral bandwidth of less than 1 pm and an electrical tuning range of 0.4 nm.

Another method makes use of the rotatory dispersion of z-cut quartz crystals. KATO and SATO (1972) used one dextro- and one levo-rotatory quartz crystal of 45 mm length each between 3 polarizers (Fig. 1.45). The tuning rate of this arrangement is 0.24 nm/degree if the

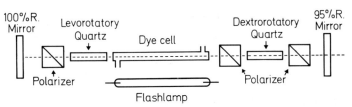

Fig. 1.45. Dye laser with double-stage rotatory dispersive filter. (From KATO and SATO, 1972)

central polarizer is rotated. The spectral half-width was rather wide, about 2 nm, since the rotatory dispersion of quartz is rather weak.

In an ingenious method, the large Faraday rotation in the vicinity of an atomic absorption line is used to lock the laser wavelength to the line (SOROKIN et al., 1969). In the experimental arrangement of Fig. 1.46, the dye laser emission obtained consisted of two doublets locked to the sodium D lines. The components of both doublets are displaced symmetrically above and below the atomic line and each has a spectral width of less than 0.1 cm^{-1}. The splitting of the laser doublets may be adjusted by varying sodium vapor pressure and magnetic field.

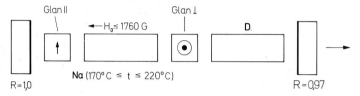

Fig. 1.46. Experimental arrangement to lock the dye laser wavelength to a spectral line. (D dye cuvette, Na sodium vapor cell, H$_0$ indicating magnetic field of solenoid surrounding cell). (From SOROKIN et al., 1969)

Dye lasers with distributed feedback might become very important active elements for integrated optics. The first distributed-feedback dye laser was described by KOGELNIK and SHANK (1971). They produced a distributed-feedback structure by inducing a periodic spatial variation of the refractive index μ according to $\mu(z) = \mu + \mu_1 \cos Kz$, where z is measured along the optic axis and $K = 2\pi/\Lambda$, Λ being the period or fringe spacing of the spatial modulation and μ_1 its amplitude. They calculated that a threshold could be reached in a dye laser with a gain of 100 and a length of 10 mm, if $\mu_1 \geq 10^{-5}$. This was easily obtained by exposing a dichromated gelatin film to the interference pattern of two coherent uv beams from a He-Cd laser. After exposure the gelatin film was developed in the usual manner and soaked in a solution of rhodamine 6 G to make the dye penetrate into the porous gelatin layer. After drying, the film was transversely pumped with the uv radiation from a nitrogen laser. Threshold was reached at pump densities of 1 MW/cm^2 and dye-laser emission less than 0.05 nm wide was observed at about 630 nm. In uniform gelatin under the same pumping conditions the emission was 5 nm wide and centered at about 590 nm.

SHANK et al. (1971) showed that a distributed-feedback amplifier can also be operated with the feedback produced by a periodic spatial variation of the gain of the dye solution. They used the experimental arrangement shown in Fig. 1.47, pumping a rhodamine 6 G solution with the fringes of two coherent beams from a frequency-doubled ruby laser, meeting at an angle 2θ. Then the wavelength of the dye laser is given by $\lambda_L = \mu_s \lambda_p / \sin\theta$, where μ_s is the index of refraction of the dye solution at the lasing wavelength λ_L, and λ_p is the pump wavelength. Thus, tuning is possible by varying either θ or μ_s. The result of angle tuning is given in Fig. 1.48, while the results of tuning by variations of the refractive index of a solvent mixture of methanol and benzyl alcohol are given in Fig. 1.49.

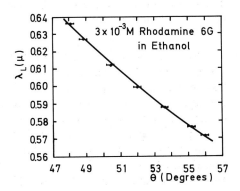

Fig. 1.47. Experimental arrangement of distributed feedback dye laser. (From SHANK et al., 1971)

Fig. 1.48. Lasing wavelength λ_L as a function of the angle θ for $3 \cdot 10^{-3}$ M rhodamine 6 G in ethanol. The points are experimental, the curve is theoretical. (From SHANK et al., 1971)

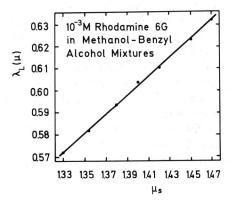

Fig. 1.49. Lasing wavelength λ_L as a function of the solvent index of refraction μ_s for $\theta = 53.8°$. The dye solution was 10^{-3} M rhodamine 6 G in a methanol-benzyl alcohol mixture. The points are experimental and the curve is theoretical. (From SHANK et al., 1971)

At about 180 kW peak pump power, the peak output power of the distributed-feedback dye laser was 36 kW. The spectral width with reduced pump power was less than 1 pm and apparently due to a single mode.

Another possibility for distributed feedback was described by KAMINOW et al. (1971). They used a sample of polymethylmethacrylate with the dimensions $4 \times 10 \times 38$ mm³, doped with $8 \cdot 10^{-6}$ mole/l of rhodamine 6 G. In this sample they produced two three-dimensional phase gratings of about $2 \times 2 \times 2$ mm by a 2-minute exposure of two intersecting uv beams of 0.7 mW each from a He-Cd laser as shown in Fig. 1.50. The sample was pumped by the second harmonic from a Nd

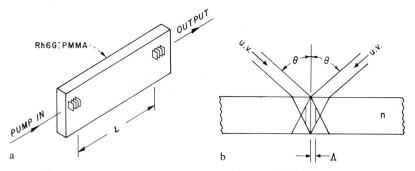

Fig. 1.50. a Dye-doped polymethylmethacrylate laser with internal grating resonator. (Dimensions: $4 \times 10 \times 38$ mm³; grating spacing $L = 20$ mm). b Preparation of gratings by intersecting uv beams through the broad face of the plastics sample. (From KAMINOW et al., 1971)

laser. Again, the dye-laser wavelength was determined by the Bragg condition $\lambda_L = 2\mu_s \Lambda$ that must be fulfilled, with $2\Lambda = \lambda_{uv}/\sin\theta$. The output consisted of one strong line of 0.5 pm near threshold. At higher pumping power, several modes separated by $c_0/2\mu_s L$ were observed, where $L = 20$ mm is the distance between the two gratings. The application of mechanical stress to the region of the gratings allowed a tuning of 1.1 nm. Yet another possibility of producing a phase grating in polymethylmethacrylate was reported by FORK et al. (1971). They dissolved 2 mmole/l rhodamine 6 G tosylate and 0.1 mole/l acridizinium ethylhexanesulfonate photodimers in methylmethacrylate and acrylic acid and polymerized the resulting solution to form a hard, transparent plastic. The sample was then cut and polished into 1 cm × 1 cm × 1 mm chips. The photodimers were first broken to a depth of 80 μm by illumination with an erase beam of 313 nm light from a mercury arc and then selectively remade in a grating pattern by two intersecting writing beams from an argon laser, in the manner described above for the other examples. The sample was then pumped by the 10-kW pulse from a neon laser focused on the sample by a cylindrical lens. The output showed a few narrow lines with a spacing determined by the $c_0/2L$ seperation for the length $L = 1$ cm of the laser. Instead of having the distributed feedback within the laser beam, one can also provide feedback for the evanescent wave and gain within the main laser beam in the dye solution adjacent to the distributed-feedback structure (HILL and WATANABE, 1972). A schematic cross-section of such a laser is shown in Fig. 1.51. In the experimental implementation of the device the cover plate was a quartz optical flat, the organic dye solution a $3 \cdot 10^{-2}$ molar solution of rhodamine 6 G in benzyl alcohol, or a mixture of benzyl and ethyl

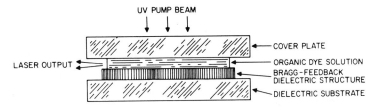

Fig. 1.51. Cross-section of the distributed feedback sidecoupled laser. (From HILL and WATANABE, 1972)

alcohols, and the feedback structure was a gelatin film grating on a glass substrate. The dichromated gelatin film had been exposed to two coherent intersecting laser beams and developed as described above. By varying the relative refractive indices of dye solution and gelatin film one could either have normal-wave gain and evanescent-wave feedback, or evanescent-wave gain and normal-wave feedback, or gelatin film and dye solution acting together as waveguide for higher-order modes for the case of nearly equal refractive indices.

1.7. Dye-Laser Amplifiers

The dye-laser oscillators discussed above are broadband amplifiers with selective or non-selective regenerative feedback. Because of its high inherent gain, a dye laser needs very little feedback to reach the threshold of oscillation. Thus, it is usually somewhat difficult to build a dye-laser amplifier, carefully avoiding all possibilities of regenerative feedback. The relatively few reports on dye-laser amplifiers will be discussed now.

The first report on broadband light amplification in organic dyes pumped by a ruby laser was by BASS and DEUTSCH (1967). They set a

Raman cell containing toluene in the ruby-laser beam and behind it a dye cell containing DTTC dissolved in DMSO. The ruby beam and the first stimulated Raman line pumped the dye solution, and broadband laser emission was obtained with the four-percent Fresnel reflection from the cuvette windows. However, when the concentration of the dye was set to a value such that the dye would lase near or at the wavelength of the second Stokes line at 806.75 nm, the broadband oscillation of the dye was quenched and the sharp Stokes line strongly amplified instead. The Raman signal being present from the beginning of the pump process used up all available inversion in the dye so that no free oscillation could start. This, too, is an experimental proof of the homogeneous broadening of the fluorescence band of the dye, at least on a nanosecond scale. Since there was four-percent feedback here, the amplification process was a multipass amplification. Similar results were obtained with CS_2 as Raman liquid and with cryptocyanine as amplifying dye. These results were confirmed by a similar investigation by DERKACHEVA and SOKOLOVSKAYA (1968).

In the ingenious experimental arrangement shown in Fig. 1.52 HÄNSCH et al. (1971 b) obtained broadband, wide-angle light amplification in several dye solutions pumped by a nitrogen laser. The amplifier cells with an active length of 1.3 mm were made from 10-mm-diameter Pyrex tube sections. The antireflection-coated windows were sealed to the ends under a wedge angle of about 10° to avoid multiple reflections. In a typical experiment two cells were used, one acting as an amplifier and the other as an oscillator. Both were filled with the same dye solution and excited simultaneously by the same nitrogen laser. Part of the dim fluorescent light of the oscillator cell, showing a noticeable preference for near-axial propagation (superradiance), was collected by a field lens and focused by an additional multielement photographic lens into the

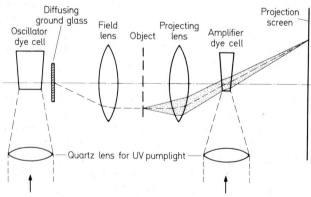

Fig. 1.52. Test set-up for image amplification using a dye laser. (From HÄNSCH et al., 1971b)

active volume of the amplifier cell. Here it was amplified and emerged as a bright light cone at the other end, illuminating a circular area on a projection screen. If now any object, such as a photographic transparency, was put into the object plane of the photographic lens, it gave rise to a bright projected image on the screen despite its own faint illumination. The gain was determined by comparing the output with and without excitation of the amplifier, absorption in the amplifier cell and background, i.e. stimulated emission of the amplifier alone, being taken into account as corrections. For small signals (input energies of up to 8 μJ), a single-pass gain of 1000 or 23 dB/mm was obtained when a rhodamine 6 G solution was excited by 100-μJ pulses of the nitrogen laser. At input signals of 25 μJ the gain dropped to 14 dB/mm, indicating saturation. With rhodamine B and fluorescein, the gain coefficients were about 4 dB/mm lower.

ERICKSON and SZABO (1971) also used a dye cell pumped by a nitrogen laser as an amplifier. The 1-cm cell was placed in a resonator consisting of a 99% and a 40% reflectivity mirror spaced 4.2 cm apart. When the acid form of 4-methylumbelliferone was pumped 20% above threshold, the spectral width of the dye-laser emission was 40 nm. Injecting the 514.5 nm line of an argon laser into the resonator caused practically the same energy to be emitted in the region around 514.5 nm in a bandwidth of only 0.16 pm, or about 4 times the width of the injected argon line. This is equivalent to a multi-pass amplification in the dye cell of 10^5 or a single-pass gain of 100. A similar regenerative amplifier experiment was described by VREHEN and BREIMER (1972). They longitudinally pumped a dye cell filled with a mixture of cresyl violet and rhodamine B with a frequency-doubled Nd laser. The cell was placed in a resonator consisting of two mirrors, one of which had 10% reflectivity at 530 nm to pass the pumping laser beam and 95% reflectivity at 632.8 nm, while the other had 20% transmission at this wavelength for the injection of 1 mW from a He-Ne laser. Here, too, the total output energy from the dye laser with and without injection was found to be constant, the effective total gain for the injected radiation being 10^6. The output of the regenerative amplifier consisted of one or several longitudinal modes centered around the injected line, the number and linewidth depending on the length of the cavity that could be tuned piezoelectrically.

The gain obtainable in flashlamp-pumped dye-laser amplifiers is much smaller than that in laser-pumped amplifiers because of the high triplet losses. HUTH (1970) measured the wavelength and time dependence of a flashlamp-pumped amplifier. The amplifier cell had a 5.3 mm inner diameter and was 3.8 cm long. It had antireflection-coated windows with 30-minute wedges. Pumping was by a flashlamp of 3-μsec half-width in an elliptical cylinder and with typically 20 J energy. The dye solution was a $2 \cdot 10^{-4}$ molar solution of rhodamine 6 G in ethanol. The signal to be amplified was derived from a dye-laser oscillator, had a bandwidth of 0.1 nm tunable by a grating over the range from 570–630 nm, and a peak power of typically 20 W/cm^2. It had a duration of about 400 nsec and could be shifted in time over the flashlamp pulse length. The maximum gain thus found was 2.3, or 95 dB/m. Even less gain was found in a six-stage amplifier chain by Flamant and Meyer (1971) who measured an energy gain of only 6.0 in the whole chain. This low value was attributable to the very high transmission losses, indicated by the fact that the ratio of the amplifier outputs when pumped and not pumped, termed "apparent gain" by the authors, was 700. These investigations suggest that great improvements in the operation of amplifiers are possible when all parameters are carefully optimized. Injection of a strong monochromatic signal into a regenerative dye-laser amplifier (termed "forced oscillator" in this work) was used by Magyar and Schneider-Muntau (1972). The amplifier cell had an inner diameter of 9 mm and a length of 160 mm and was pumped in a close-coupled configuration by six linear air-filled flashlamps, enclosed by a cylindrical reflector of aluminum foil. The energy of the pump pulse was 4.2 kJ, its risetime 2 μsec and its half-width 15 μsec. The dye was rhodamine 6 G in water of somewhat less than $7 \cdot 10^{-5}$ molar concentration, with the addition of 1.5% Ammonyx and 0.2% cyclooctatetraene. One resonator mirror was 99% reflecting, the outcoupling mirror 50%. The output without injection was 1.6 J in a 12 nm wide spectral band. A cell could be placed into the cavity under the Brewster angle that had the twofold purpose of containing an absorbing dye solution and injecting the signal by reflection from its front window. The injected signal was derived from a dye laser oscillator that was spectrally narrowed and tuned by two tilted Fabry-Perot interferometers and had a maximum output of 55 mJ in 400 nsec and a bandwidth of less than 10 pm. Careful timing of the signal resulted in most of the energy of the amplifier appearing in the amplified injected line. Complete frequency locking, however, could only be achieved by adding a few drops of a solution of a suitable absorbing dye, e.g. 1,1′-diethyl-4,4′-cyanine iodide, to the absorber cell which previously contained only solvent. The output then contained only the injected line and had a total energy of 600 mJ. This frequency locking by an absorbing dye was first demonstrated with ruby lasers (Opower and Kaiser, 1966). Since the injected signal had a duration of only 400 nsec while the amplified signal was 5 μsec long, only the front part was true amplification and the forced oscillator continued its emission at the same frequency in the later part of the pulse. Because of the complexity of the operation of this device, it is difficult to make a meaningful statement concerning power or energy gain. Nevertheless, the practical interest of this type of regenerative amplifier or forced oscillator is considerable.

1.8. Outlook

To end this chapter, the reader might be interested in a few speculative remarks about possible trends for future developments of dye lasers. Regarding the chemical aspects, the reader is referred to the concluding remarks in Chapter 4. The most important topics with regard to the physical aspects of dye lasers are new pumping methods and pump-light sources, followed closely by the physical dimensions of dye lasers. Some properties can be extrapolated to foreseeable physical limits.

The usual pumping methods using lasers and flashlamps will certainly be improved with respect to efficiency, power, control over pulse shape, and several other parameters. Before an incoherently pumped cw dye laser can be realized, new high-power arc lamps will have to be developed. The spectrum of such lamps must be well matched to the absorption spectrum of the dye and have a very high brightness in order to reach the high values of absorbed power per cm^3, given by SNAVELY (1969).

For very small pulsed or continuous dye lasers direct electrical pumping might be feasible. Some dyes are reported to form relatively stable anions and cations in certain aprotic solvents, e.g. the ions of diphenylanthracene (DPA) in dimethylformamide, which can be formed electrochemically and which, at recombination, leave one molecule in the excited state so that the fluorescence of the neutral molecule is emitted (CHANDROSS and VISCO, 1964; HERCULES, 1964):

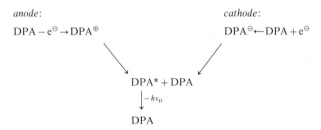

This method would have definite advantages for applications with integrated optics.

For very large volumes of dye solution, flashlamp-pumping would be very cumbersome and expensive; in this case chemical energy storage is more economical, one kilogram of explosive storing about 5 MJ of mechanical energy. This high energy content can be used to excite a shock wave in a gas (so-called argon bombs) (HELD, 1968) which thus becomes brightly luminescent. A cross-section through a possible structure utilizing such an argon bomb is shown in Fig. 1.53 (SCHÄFER, 1969). The cylindrical mantle of explosive is ignited simultaneously at

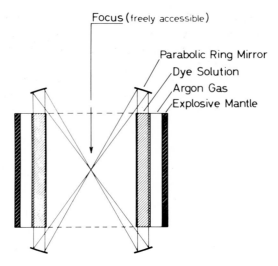

Focus (freely accessible)

Parabolic Ring Mirror
Dye Solution
Argon Gas
Explosive Mantle

Fig. 1.53. Cross-section of a dye laser pumped by an argon bomb. (From SCHÄFER, 1969)

many lines of the circumference and excites a compressive shock wave in the argon layer that is traveling towards the symmetry axis of the structure. The dye solution is excited by the luminescent output from the shock wave and the concomitant superradiant output can be focused, e.g. by a parabolic ring mirror to the center point of the axis. One can expect outputs of several kJ focused on this spot, which would be useful for plasma experiments.

An extrapolation of present dye-laser properties into the next few years would give an estimate of extended wavelength coverage of 250 to 1500 nm for pulsed lasers and 380 to 950 nm for cw lasers. Together with frequency mixing and multiplication and stimulated Raman emission, this would in effect give complete coverage from the vacuum ultraviolet to the far-infrared.

The maximum power output of a laser is reached when a pulse containing the saturation energy (in photons per square cm) $E_s = 1/\sigma_{fl}$ is being amplified. For a dye $\sigma_{fl} \approx 10^{-16}\,\mathrm{cm}^2$, and thus $E_s = 10^{16}$ photons/cm^2, equivalent (at 600 nm) to a power of 330 MW/cm^2 for a pulse of 10 psec duration. Since the cross-section of a dye solution is practically unlimited, pulses of many gigawatts peak power could be generated with dye lasers.

The pulse energies of more than 1 J/cm^3 of dye solution that have been reported are higher than the stored energy and can only be obtained by multiple pumping of the dye molecules by a strong pump-light pulse. The stored energy at 10% inversion, 600 nm laser wavelength and a concentration of 10^{-3} mole/l is only $20\,\text{mJ/cm}^3$. Since this energy is stored for typically a few nsec, the molecules can be pumped many times during a pulse of some μsec duration. Thus, multikilojoule pulses from a few liters of dye solution appear feasible provided the problem of sufficiently strong pumping of such large volumes is resolved.

For a discussion of ultimate frequency stability of cw dye lasers, the reader is referred to Chapter 2, while the question of the shortest possible pulses is discussed in Chapter 3.

2. Continuous-Wave Dye Lasers

B. B. SNAVELY*

With 18 Figures

The operation of the continuous-wave (cw) dye laser is based upon the same molecular states as that of pulsed lasers. However, some loss mechanisms that are relatively unimportant in the pulsed laser tend to dominate the performance of the cw laser. For example, the accumulation of molecules in the triplet state plays a major role in determining the efficiency of the cw laser, whereas the triplet state is relatively unimportant in pulsed lasers. Also, optical inhomogeneities produced in the active medium by heating resulting from excitation must be carefully controlled for the laser to operate continuously. Excitation sources for pulsed lasers are generally capable of producing intensities greatly in excess of that required to reach laser threshold. Pump sources for cw lasers, on the other hand, are often marginal. For these reasons, the efficient operation of a cw dye laser requires careful design and construction to minimize extraneous optical losses. The mechanical and optical tolerances are generally much more severe than those for pulsed dye lasers. In this chapter the analysis and design of cw dye lasers will be considered. Tuning systems will be discussed and the characteristics of some experimental cw laser systems reviewed.

As with pulsed dye lasers, the cw laser consists of an optically excited fluorescent dye solution within a suitable resonator. The mechanism for the production of stimulated emission in the dye solution is reviewed in Fig. 2.1.

Optical gain is associated with stimulated transitions $a \leftarrow B$ between the states labeled S_1 and S_0 of the singlet-state manifold. The population of S_1 is obtained by optical excitation at a wavelength corresponding to the transition $A \rightarrow b$. In some cases the excitation process may correspond to $A \rightarrow b'$, shown by the dashed extension of the $A \rightarrow b$ transition, though this is generally unfavorable owing to the large amount of energy that must be dissipated as heat in the subsequent relaxation to the upper laser level, $S_1 \leftarrow S_2$.

Molecules in the upper laser level, B, may decay by the competing processes $S_0 \leftarrow S_1$ or $T_1 \leftarrow S_1$. T_1 is the lowest of a manifold of triplet states. The transition $S_0 \leftarrow S_1$ may occur either by spontaneous decay,

* This work was carried out when the author was with the Research Laboratories of Eastman Kodak Company, Rochester, N.Y. 14650/USA.

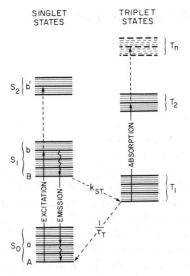

Fig. 2.1. Schematic energy level diagram for a dye molecule. The processes of importance in the operation of the cw dye laser are indicated by arrows. The heavy horizontal lines represent vibrational sublevels of the electronic states and the lighter lines represent levels due to interaction of the dye molecule with a solvent

which represents a loss, or by stimulated emission, the desired result. The relative transition rates for these processes are one factor that determines the operating efficiency of the laser. Any nonradiative $S_0 \leftarrow S_1$ decay also lowers the efficiency of the laser. This process, termed *internal conversion*, is generally negligible for dyes capable of lasing continuously.

An essential difference between the operation of the cw laser and that of pulsed lasers is the importance of the $T_1 \leftarrow S_1$ process. Although the probability for this process is relatively small, in good laser dyes, several percent of the excited molecules may make the transition. Molecules which arrive in T_1 remain there for a time that is long compared to the characteristic times involved in the lasing process. The transition $S_0 \leftarrow T_1$ is forbidden and the rate constant for this process ranges from 10^3 to 10^7 sec^{-1}, depending upon the environment of the dye molecule. The lifetime may be shortened if molecules, such as O_2, which tend to enhance the $S_0 \leftarrow T_1$ (SNAVELY and SCHÄFER, 1969; PAPPALARDO et al., 1970) process are nearby and, conversely, it may be quite long if there are no triplet quenching molecules in the dye solution.

The transition $S_0 \leftarrow T_1$ is, in all cases, slow with respect to $S_0 \leftarrow S_1$ which has a rate constant for spontaneous decay of about 2×10^8 sec^{-1}. The state T_1 thus acts as a trap and, for high excitation rates, a non-negligible fraction of the total number of dye molecules may reside in T_1.

Of course, if N_T, the population of T_1, approaches N, the total dye concentration, lasing is not possible because there are no molecules available in the $S_0 \leftarrow S_1$ lasing path.

Lasing may also be quenched when $N_T \ll N$. T_1 is the lowest of a manifold of states of which the next higher state T_2 is shown. Transitions between the states of this manifold, $T_1 \rightarrow T_2$ for example, are allowed since no change in electron spin is required. The T_1 molecules have a large absorption cross section for the $T_1 \rightarrow T_2$ and $T_1 \rightarrow T_n$ processes shown in Fig. 2.1. Unfortunately, the absorption spectrum associated with T_1 generally overlaps the $S_0 \leftarrow S_1$ fluorescence of a dye molecule. Thus, it produces an optical loss at wavelengths for which optical gain is to be produced. This loss must be overcome by an increase in excitation power with respect to the power required if the loss were not present. At the least, the efficiency of the lasing process suffers as a result. If the absorption loss is too large, lasing may be prevented altogether.

The strength of an absorption process may be expressed in terms of the *molecular absorption cross section* $\sigma(\lambda)$ defined by the relation

$$I(\lambda, d) = I(\lambda, 0) \, e^{-N\sigma(\lambda)d} \, .$$

Here $I(\lambda, 0)$ is the intensity of a light beam incident upon an absorbing sample of thickness d with a concentration of N absorbing molecules per cm^3. $I(\lambda, d)$ is the intensity of the transmitted beam. The absorption cross section has the dimensions of cm^2 and is effectively the area of an absorbing molecule at the wavelength λ.

Absorption cross sections for various processes in the laser dye rhodamine 6G are shown in Fig. 2.2. The cross section for $S_0 \rightarrow S_1$, labeled $\sigma_s(\lambda)$, is presented for a water solution of the dye. The long wavelength tail of $\sigma_s(\lambda)$ is also shown with an expanded ordinate. The water solution is often used in cw dye lasers (PETERSON et al., 1970). The fluorescence spectrum $E(\lambda)$, corresponding to the spontaneous $S_0 \leftarrow S_1$ process and the stimulated emission cross section $\sigma_{em}(\lambda)$ derived from $E(\lambda)$, are shown for comparison with $\sigma_s(\lambda)$. The derivation of $\sigma_{em}(\lambda)$ will be discussed in the following section.

The absorption spectrum for $T_1 \rightarrow T_n$, $\sigma_T(\lambda)$, is also shown in Fig. 2.2. This spectrum was measured by MORROW and QUINN (1973) for an ethanol solution of rhodamine 6G. Unfortunately, the $\sigma_T(\lambda)$ spectrum has not been measured for a water solution. From the comparison of triplet spectra for different solvents with other dye molecules it is not expected that the triplet spectrum for rhodamine 6G in water would differ greatly from that shown.

The data of Fig. 2.2, along with a knowledge of the rate constants for the processes shown in Fig. 2.1., may be used to deduce the longest laser pulse which can be obtained from this solution if the triplet state

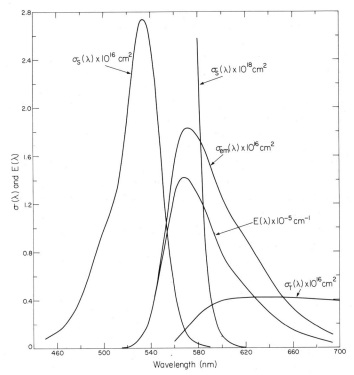

Fig. 2.2. Spectrophotometric data for the dye rhodamine 6G. The singlet absorption, $\sigma_S(\lambda)$, and emission, $E(\lambda)$, were obtained from a solution of the dye in water plus 2% Ammonyx LO, a commercial surfactant. The triplet-state spectrum, $\sigma_T(\lambda)$, was obtained by MORROW and QUINN (1973) with an ethanol solution. The dye concentration in all cases was 10^{-4} M. $E(\lambda)$ has been normalized so that $\int E(\lambda)\,d\lambda = 0.92$, the measured quantum yield for fluorescence

population is uncontrolled. That is, the longest laser pulse duration t_{max} consistent with a triplet lifetime τ_T which is very large with respect to τ, the lifetime for the spontaneous $S_0 \leftarrow S_1$ process.

If the rate constant for the $T_1 \leftarrow S_1$ process is $k_{ST}(\sec^{-1})$, the rate at which molecules enter T_1 is given by $N_{1C}k_{ST}$ where, N_{1C} is the *critical inversion*, the population of S_1 required for the gain of the active medium to balance the intrinsic optical losses. Above laser threshold the population S_1 is fixed at N_{1C}; it does not increase significantly with excitation (SMITH and SOROKIN, 1966). The population of T_1, N_T, will then be governed by the rate equation:

$$\frac{dN_T}{dt} = N_{1C}k_{ST} - \frac{N_T}{\tau_T}. \tag{2.1}$$

If it is assumed that the critical inversion is produced at time $t = 0$, and if $\tau_T \to \infty$,

$$N_T(t) = N_{1C}k_{ST}t .\tag{2.2}$$

It is necessary at this point to anticipate a result from the analysis in the following section. If the molecular cross section for stimulated emission is $\sigma_{em}(\lambda)$ and the cross section for triplet-triplet absorption is $\sigma_T(\lambda)$ the intrinsic gain will just balance the loss due to the triplet state when

$$N_{1C}\sigma_{em}(\lambda) = N_T\sigma_T(\lambda) ,$$

where N_1 is the molecular concentration in the state S_1. From Fig. 2.2 it is seen that $\sigma_T(\lambda) \sim \sigma_{em}(\lambda)/10$ at the peak of $\sigma_{em}(\lambda)$. For the corresponding wavelength the net gain vanishes when $N_T \approx 10 N_{1C}$. From (2.2), therefore,

$$t_{max} \approx 10/k_{ST} .$$

The quantity k_{ST} can be estimated from a knowledge of ϕ, the fluorescence quantum efficiency, and τ by means of the relation

$$k_{ST} = \tau^{-1}(1 - \varphi) .$$

For $\tau = 6 \times 10^{-9}$ sec and $\phi = 0.92$[1], as is approximately true for rhodamine 6G, $k_{ST} = 1.6 \times 10^7$ sec^{-1} in which $t_{max} \sim 6 \times 10^{-7}$ sec. This is consistent with the observation of the laser pulse duration for systems in which no control is exercised over the population of T_1 (SNAVELY, 1969; SOROKIN et al, 1968).

From the foregoing discussion it is apparent that the population of T_1 must be limited if the cw operation of a dye laser is to be achieved. This can be accomplished by the control of k_{ST} or τ_T. The maximum allowable value for k_{ST} is readily estimated by an extension of the reasoning presented above. In the steady state, $dN_T/dt = 0$ and (2.1) yields for the equilibrium population of T_1

$$N_T = N_{1C}k_{ST}\tau_T .\tag{2.3}$$

Again assuming that the intrinsic gain vanishes when $N_T/N_{1C} = \sigma_{em}/\sigma_T = k_{ST}\tau_T$, a maximum value for $k_{ST}\tau_T$ of ≈ 10 is obtained from the data of Fig. 2.2. If cw operation is to be achieved, some means of quenching the triplet state concentration rapidly enough that $k_{ST}\tau_T < 10$ must be provided. This may be achieved by chemical additives or by rapid flow of the dye through the excited region.

---

[1] Measured by F. GRUM, Eastman Kodak Research Laboratories.

2.1. Gain Analysis of the cw Dye Laser

2.1.1. Analysis at Laser Threshold

Analysis of the triplet state kinetics on the basis of the triplet state rate equation has provided a useful insight into the importance of the control of the triplet state population. Many other aspects of the performance of the dye laser can be understood from an analysis in terms of rate equations. In the treatment which follows the rate equation approach developed by SNAVELY (1969) and by PETERSON et al. (1971) will be followed for the description of the laser at threshold. The effects of system parameters upon tuning and threshold will be examined.

A system of the form shown in Fig. 2.3, consisting of a longitudinally excited region within a two element optical resonator will be considered. The resonator is hemispherical. It is assumed that the concentration of molecules in S_1 is N_1 cm^{-3} and is uniform throughout the active volume. This assumption will be modified when the behavior of the laser above threshold is considered. For the moment, however, it is not restrictive.

If $I(\lambda, z)$ is the intensity of the lasing mode the net rate at which $I(\lambda, z)$ increases along the axis of the laser cavity, labeled as the z direction, will be given by

$$\left(\frac{dI(\lambda, z)}{dz}\right)_{\text{total}} = \left(\frac{dI(\lambda, z)}{dz}\right)_{\text{stim}} - \left(\frac{dI(\lambda, z)}{dz}\right)_{\text{sing.}} - \left(\frac{dI(\lambda, z)}{dz}\right)_{\text{trip}} \quad (2.4)$$

where $\left(\dfrac{dI(\lambda, z)}{dz}\right)_{\text{stim}}$ is the rate at which the intensity increases due to stimulated emission, and $-\left(\dfrac{dI(\lambda, z)}{dz}\right)_{\text{sing.}}$ and $-\left(\dfrac{dI(\lambda, z)}{dz}\right)_{\text{trip}}$ are the rates at which it decreases due to $S_1 \rightarrow S_2$ and $T_1 \rightarrow T_n$ processes, respectively. In the discussion which follows the explicit functional dependence of quantities, such as $I(\lambda, z)$, will not be indicated when ambiguity is unlikely to result. For instance, "$I(\lambda, z)$" will be written simply as "I".

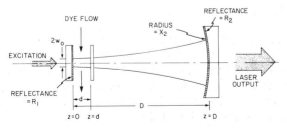

Fig. 2.3. Geometry of the cw laser considered in the analysis

Expressions for the terms of (2.4) have been derived by many authors and will be taken from the literature. To obtain $\left(\dfrac{dI}{dz}\right)_{\text{stim}}$ use is made of the identity $\left(\dfrac{dI}{dz}\right) = \left(\dfrac{dI}{dt}\right)\left(\dfrac{dz}{dt}\right)^{-1} = \dfrac{n}{c}\dfrac{dI}{dt}$ in which case (YARIV, 1967) $\left(\dfrac{dI}{dz}\right)_{\text{stim}} = \dfrac{N_1 \lambda^4 E(\lambda)}{8\pi\tau c n^2} I = \sigma_{\text{em}}(\lambda) I$. The quantities n and c are the refractive index and the velocity of light, respectively. The quantity $\sigma_{\text{em}}(\lambda)$ as defined by this equation is plotted in Fig. 2.2. $E(\lambda)$ is the spontaneous emission lineshape function normalized so that $\int\limits_0^\infty E(\lambda)\, d\lambda = \phi$, the fluorescence quantum yield, and τ is the observed fluorescence decay time for spontaneous emission. The quantities $(dI/dz)_{\text{sing.}}$ and $(dI/dz)_{\text{trip.}}$ have the form $(dI/dz) = -I N_\alpha \sigma_\alpha(\lambda)$ where $\sigma_\alpha(\lambda)$ is an absorption cross section. The absorption cross section for $S_0 \rightarrow S_1$ process has previously been denoted as $\sigma_s(\lambda)$ and that for $T_1 \rightarrow T_2$ processes as $\sigma_T(\lambda)$. N_α is the appropriate molecular density, the concentration of molecules in S_0, which is N_0, or N_T.

Using these relationships the rate equation for photon production becomes

$$\left(\dfrac{dI}{dz}\right)_{\text{total}} = I\left\{N_1 \sigma_{\text{em}}(\lambda) - N_0 \sigma_s(\lambda) - N_T \sigma_T(\lambda)\right\}. \qquad (2.5)$$

Defining the gain of the active medium as

$$g(\lambda) = \dfrac{1}{I}\left(\dfrac{dI}{dz}\right)_{\text{total}}$$

and integrating (2.5) over a round trip through the active medium, accounting for the finite reflectances $R_1(\lambda)$ and $R_2(\lambda)$ of the mirrors, yields $G(\lambda)$, the round trip gain,

$$G(\lambda) = \int\limits_{\text{round trip.}} g(\lambda)\, dz = N_1 \sigma_{\text{em}}(\lambda)\, 2d - N_0 \sigma_s(\lambda)\, 2d - N_T \sigma_T(\lambda)\, 2d$$
$$- \ln\left[R_1(\lambda)\, R_2(\lambda)\right]. \qquad (2.6)$$

At laser threshold $G(\lambda) = 0$ so that

$$N_1 \sigma_{\text{em}}(\lambda) - N_0 \sigma_s(\lambda) - N_T \sigma_T(\lambda) - \dfrac{1}{2d}\ln\left[R_1(\lambda)\, R_2(\lambda)\right] = 0.$$

In the steady state the triplet and singlet state concentrations are related by (2.3) which expresses the *equilibrium triplet approximation* utilized by PETERSON, et al. [1971]. Using this relationship and the condition $N = N_0 + N_1 + N_T$, the threshold condition can be written as

$$\{\sigma_{em}(\lambda) + \sigma_s(\lambda) + k_{ST}\tau_T\,[\sigma_s(\lambda) - \sigma_T(\lambda)]\}\,N_{1C} - \sigma_s(\lambda)N + r(\lambda) = 0 \qquad (2.7)$$

where $r(\lambda) = -(1/2d)\ln[R_1(\lambda)R_2(\lambda)]$. The term in brackets is a growth coefficient relating the net rate at which photons are produced to the population of the upper laser level. It depends only upon the spectro-photometric parameters of the dye solution shown in Fig. 2.2 and is independent of laser system parameters which are contained in r.

If the growth coefficient in brackets is written as $\gamma(\lambda)$, (2.7) can be recast to solve for the critical inversion N_{1C} as a function of wavelength and system parameters, i.e.

$$\frac{N_{1C}}{N} = \frac{1}{\gamma(\lambda)}\left[\sigma_s(\lambda) + \frac{r(\lambda)}{N}\right]. \qquad (2.8)$$

Equation (2.8) describes a critical inversion surface. As demonstrated by PETERSON et al. (1971), examination of this surface provides insight into the operation of the cw dye laser, especially its tuning behavior.

The critical inversion surface has its *intrinsic* or *self-tuned* form when $r(\lambda)$ is nondispersive, i.e. the mirror reflectance is independent of wave-length. The self-tuned critical inversion surface for the dye rhodamine 6G is shown in Fig. 2.4. In plotting this surface a value for $k_{ST}\tau_T = 0.9$, as found by SIEGMAN et al. (1972) has been used. The value for τ of 6×10^{-9} sec has been taken from the work of SCHÄFER et al. (1973). Other parameters have been taken from the data of Fig. 2.2.

Without excitation $N_1/N = 0$. As the excitation intensity is increased, N_1/N increases and at some time intersects the critical inversion surface. When this occurs oscillation begins and the population of N_1 is clamped at N_{1C}. Thus, the critical inversion is constrained to lie along the valley of the surface as indicated by the dashed line in Fig. 2.4. The range over which the wavelength can be tuned by adjustment of mirror reflectance and dye concentration is found from the extremes of the dashed curve.

The self-tuning range for the rhodamine 6G laser is seen to extend from approximately 560–650 nm. At the short wavelength end of the tuning range N_{1C}/N approaches unity while at the long wavelength limit the cross section for stimulated emission becomes too small to overcome the loss due to singlet state and triplet state absorption.

Tuning of a system over the self-tuned range is accomplished by adjusting the ratio r/N. From Fig. 2.4 it is seen that increasing r, while keeping N fixed, causes the laser to oscillate at shorter wavelengths, while

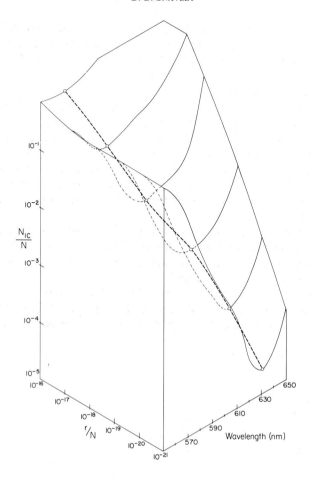

Fig. 2.4 Intrinsic critical inversion surface resulting from a plot of (2.8) for the dye rhodamine 6G in water plus 2% Ammonyx LO. The data of Fig. 2.2 have been used in the analysis. The fluorescence decay time τ was taken as 6.1×10^{-9} sec, as measured by Schäfer et al. (1973). With nondispersive mirrors forming the resonator, the laser is constrained to operate along the dashed curve

increasing N, keeping r fixed, increases the laser wavelength. This behavior is consistent with experimental results (Sorokin et al., 1968; Schäfer et al., 1966).

In practice, the useful tuning range is less than that indicated above. It is generally not possible to provide an excitation intensity sufficient to produce an inversion greater than $N_{1c}/N \approx 0.1$, which places a limitation upon the shortest wavelengths that may be attained. The minimum value of cavity loss due to scattering in the active medium and diffraction loss of the resonator determines the longest wavelength for oscillation. These losses have not been included in the rate equation analysis. For a carefully designed system they are generally much smaller than the loss due to triplet state absorption.

Of course, the laser may be operated at points other than those along the dashed line by using dispersive mirrors to form the optical resonator. A laser operated in this way is termed *extrinsically* tuned. The inclusion of dispersive optical elements within the cavity actually modifies the shape of the critical inversion surface. The modified surface can never lie lower than the intrinsic surface, however. Discussion of the surface shape under these conditions will be deferred to the section on tuning of the cw dye laser.

From the critical inversion surface it is possible to estimate the excitation power required to reach laser threshold for a cw device. The dependence of the excitation power upon some of the laser system parameters can also be determined.

Projection of the dotted curve of Fig. 2.4 onto the N_{1C}/N vs. r/N plane gives a representation of the critical inversion as a function of the extrinsic loss. Since the dotted path is the locus of points for which

$$\frac{d}{d\lambda}\left(\frac{N_{1C}}{N}\right) = 0 \text{ the condition}$$

$$\frac{\gamma(\lambda)}{\gamma'(\lambda)} = \frac{\left[\sigma_s'(\lambda) + \dfrac{r'(\lambda)}{N}\right]}{\left[\sigma_s(\lambda) + \dfrac{r(\lambda)}{N}\right]} = \frac{\sigma_s'(\lambda)}{\sigma_s(\lambda) + \dfrac{r(\lambda)}{N}},$$

where prime denotes the wavelength derivative, holds. This condition is satisfied at a particular wavelength for each value of r/N. Substitution of the wavelength found for a given r/N into (2.8) then yields N_{1C}/N for that value. Values of N_{1C}/N vs. r/N found in this way are plotted in Fig. 2.5 for rhodamine 6G. In these curves $\mu = k_{ST}\tau_T$ is chosen as a parameter.

Above the low r transition region the value of N_{1C}/N is relatively independent of μ. PETERSON et al. [1971] have shown that in this region the critical inversion is given approximately by

$$\frac{N_{1C}}{N} = 5.8 \times 10^{11}\left(\frac{r}{N}\right)^{3/4}. \tag{2.9}$$

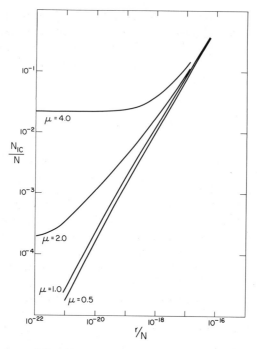

Fig. 2.5. Dependence of the critical inversion upon the ratio r/N for nondispersive resonator mirrors. Curves for various values of the parameter $\mu = k_{ST}\,\tau_T$ are shown

The total excitation power required is proportional to the product $N_{1c}d$. From (2.9) it is found that

$$P_p \propto N_{1c}d = 5.8 \times 10^{11} \left(\frac{\ln R_2}{2} \right) d^{1/4} ,$$

where it has been assumed that $R_1 = 1.0$. Thus, to minimize the excitation power the active length should be kept small.

From Fig. 2.5 it is seen that for the higher values of $\mu, N_{1c}/N$ approaches a constant, independent of r/N, for small values of r/N. For any given value of μ, there is a practical minimum for r/N. Reduction of the extrinsic losses below this value does not significantly reduce the amount of excitation power required to reach threshold. There is also a limit imposed upon r/N by practical considerations. To reach the saturation region of the $\mu = 0.5$ curve for rhodamine 6 G requires $r/N = 10^{-22}$. Values of N much greater than 6×10^{17} cm^{-3} (10^{-3} M) are not useful because the fluorescence quantum yield of the dye solution begins to drop at such high concentrations. The minimum value required for r is, therefore, approximately 6×10^{-5} requiring R_2 to be about 99.999 %, an unrealistically high value. With mirrors of the highest quality 99.9 % is attainable, yielding $r/N = 10^{-20}$. With this value of r/N the laser is operating well out of the saturation region for small values of μ. The critical inversion is then relatively independent of μ.

The minimum excitation power density required to reach laser threshold, P_p/A, can be found directly from Fig. 2.5 since

$$\frac{P_p}{A} \cong \frac{N_1 c h v_p d}{\tau} \text{ watts/cm}^2 .$$

This is the excitation power at the frequency v_p per unit area of the lasing cross section in the longitudinally excited geometry of Fig. 2.3. For the dye rhodamine 6G with $\mu = 1$, an active length of 0.1 cm, and $R_2 = 0.995$, the calculated power density is approximately 3.5 kW/cm² at 530 nm.

The threshold excitation for practical systems is usually about 50 kW/cm². A minimum power density of 100 kW/cm² should be used in the excitation of a dye laser in order to provide a reasonable operating efficiency. With lower excitation power the dye laser operates close to the threshold and the overall efficiency is low. To obtain the necessary power density it is most attractive to use a cw gas laser for the excitation. The flux required to excite a yellow or red dye can be approached at the surface of a high pressure mercury arc lamp. However, the losses introduced by the optics required to transfer energy from the source to the active region are so great that available high pressure arcs are not attractive as excitation sources for cw dye lasers. No results have yet been published on the attempts to develop arc lamps suitable for the excitation of cw dye lasers. Of the available lasers operating in the visible region of the spectrum, the argon ion laser is capable of the highest output power. Commercial lasers with single or multiple line output powers of several watts are suitable for the excitation of a cw dye laser.

Given the excitation power of one to twenty watts, the beam diameter at the waist of the TEM_{00} mode of the dye laser is fixed at 10–20 μm. The small mode diameter places restrictions upon the design of the laser resonator. To minimize diffraction losses a spherical or confocal cavity is chosen. The confocal parameter (KOGELNIK and LI, 1966), or optical length of the cavity, must be small to provide a beam waist small enough to match the excitation beam. For most cw dye lasers the confocal parameter is several millimeters.

The spherical, or hemispherical resonator is, in practice, more convenient than a confocal resonator in that the physical length of the spherical resonator is twice as long as that of the confocal resonator, for a given confocal parameter. This simplifies the fabrication of the laser since the dimensions of the optical elements are not small. Simple plane parallel resonators are not practical for the cw dye laser since the diffraction loss is large at the small Fresnel number required.

2.1.2. Gain Analysis above Laser Threshold

In the preceding subsection it was assumed that the concentration of molecules in the upper laser level, N_1, was uniform throughout the active volume of the laser. Although the assumption simplified the derivation of the critical inversion and the threshold excitation power, as well as permitting the display of the self-tuning range, it does not correspond to an operating, longitudinally excited cw laser. For the practical device the population of the upper laser level is position dependent. It has the form $N_1(r,z)$ where r is the radial distance from the cavity axis. When the position dependence of N_1 is incorporated the gain equation is not as simple as (2.6).

To develop a gain equation which can be used in the design of cw systems the active region of the laser, as shown in Fig. 2.3, is assumed to have the geometry shown in Fig. 2.6. The hemispherical resonator has a confocal parameter b_L, which determines the waist radius of the laser beam, w_0. Laser and excitation beams are assumed to be of the TEM$_{00}$ mode. The confocal parameter for the excitation beam will be taken as b_p with the corresponding waist diameter w_{p0}. As drawn, the excitation and laser beam waists coincide at $z = 0$.

Following the analysis developed by Pike (1971) the optical gain per unit length for a beam of intensity $I_L^+(r,z)$ traveling to the right and $I_L^-(r,z)$ traveling to the left can be expressed in terms of the growth coefficient $\gamma(\lambda)$, defined in connection with (2.6) as

$$dI_L^+(r,z) = I_L^+(r,z)\,N_1(r,z)\,\gamma(\lambda)\,dz$$

and

$$dI_L^-(r,z) = I_L^-(r,z)\,N_1(r,z)\,\gamma(\lambda)\,dz\,.$$

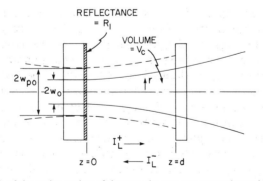

Fig. 2.6. Details of the active region of the two-element resonator shown in Fig. 2.3

If the mirror at the right of Fig. 2.3 returns a fraction R_2 of the beam $I_L^+(r,d)$ to the active region, the on-axis round-trip gain of the system will be

$$G_0 = (1 + R_2) \int_0^d N_1(0,z) \, \gamma(\lambda) \, dz . \qquad (2.10)$$

The excitation intensity $I_p(r,z)$ determines the population of the upper laser level by $N_1(r,z) = I_p(r,z) N_0 \sigma_p \tau$ where σ_p denotes the absorption cross section $\sigma_S(\lambda_p)$ for excitation radiation. The excitation irradiance for a TEM_{00} excitation beam will be

$$I_p(r,z) = \frac{4 P_p(0)}{b_p h c} \left(\frac{w_{p0}}{w_p} \right)^2 \exp\left(-2r^2/w_p^2 \right) \quad \text{photons} \quad cm^{-2} \quad sec^{-1} \quad \text{with}$$

$w_{p0}^2 = \dfrac{\lambda_p b_p}{2\pi}$ and $w_p^2(z) = w_{p0}^2 \left(1 + \dfrac{4z^2}{b_p^2} \right)$. $P_p(0)$ is the incident power in watts.

From (2.10) it is clear that the round trip gain is maximized when the single pass gain is largest. Therefore only the single pass gain needs to be considered to determine the effects of changes in parameters upon gain. Making use of the dependence of pump irradiance upon distance as given above, the on-axis single pass gain can be written as

$$G_0 = \frac{4\gamma\tau P_p(0)}{b_p h c} \int_0^d \frac{N_0 \sigma_p e^{-N_0 \sigma_p z}}{1 + 4z^2/b_p^2} \, dz .$$

PIKE [1971] has shown that G_0 may be written in the useful form

$$G_0 = \frac{4\gamma\tau P_p(0)}{b_p h c} F(u,v) \qquad (2.11)$$

where $u = N_0 \sigma_p d$, $v = N_0 \sigma_p b_p/2$ and $F(u,v) = v^2 \int_0^u \dfrac{e^{-x} dx}{v^2 + x^2}$.

$F(u,v)$ has been expressed in terms of an exponential integral which attains its maximum value of unity when u and v are very large. This occurs when $1/(N_0 \sigma_p)$, the absorption length in the dye is small in comparison with either the cell length, d, or the confocal parameter b_p. The primary conclusion to be drawn from (2.11) is that the front surface of the active medium should be placed at the waist of the excitation beam to obtain the maximum gain with a given excitation power. This is not too surprising when it is realized that the attenuation length in the dye, under the present assumptions, is independent of the excitation irradiance, whereas the gain is proportional to the irradiance. If the excitation beam is absorbed in a region of low irradiance, the gain of the system will be low as predicted by the analytical result.

The above analysis has been applied to a cw dye laser which is external to the cavity of the exciting laser. In some systems of interest the dye laser is inside the cavity of the exciting laser. In this case the excitation beam may be recirculated to provide a more uniform population of the active medium than in the case just considered. For this arrangement a low concentration of dye would be required, in order that the gain of the exciting laser would not be quenched. Although the dye concentration may be low, the upper laser level population must be comparable to that of the high concentration device. This is accomplished by the high excitation flux within the exciting laser cavity. The geometry of the system to be considered is shown in Fig. 2.7.

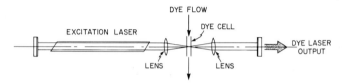

Fig. 2.7. Geometry of a cw dye laser internal to the cavity of the excitation laser

Assume that the intracavity power of the exciting laser, flowing in the $+z$ direction, P_{1C}^+, is the power incident upon the left side of the dye cell in Fig. 2.6. The excitation power within the dye cell will then be given by

$$P_p(z) = P_{1C}^+(0) \, e^{-N_0 \sigma_p z} + P_{1C}^+(0) \, e^{-N_0 \sigma_p d} \, e^{-N_0 \sigma_p (d-z)}$$
$$= 2 P_{1C}^+(0) \, e^{-N_0 \sigma_p d} \cosh\left[N_0 \sigma_p (d-z)\right].$$

This expression may be inserted into (2.11) to find the upper laser level population as a function of position along the axis. The on-axis gain is then calculated in the same manner as was done previously to yield

$$G_0 = \frac{8 \gamma \tau P_{1C}^+(0)}{b_p hc} \, e^{-N_0 \sigma_p d} \sinh(N_0 \sigma_p d).$$

Since it has been assumed for this case that the absorption length $1/N_0 \sigma_p \gg d$ the gain may be expressed as

$$G_0 \approx \frac{8 \gamma \tau P_{1C}^+(0)}{b_p hc} N_0 \sigma_p d. \qquad (2.12)$$

To compare this result with the gain of the externally excited dye laser it is assumed that the exciting laser is operating under conditions which produce the maximum N_{1C} in both cases. For the first case this occurs when $P_p(0)$ is a maximum. The maximum output power for a laser is

$$P_p(0) \cong 2\Gamma_1 P_{1C}^+(0),$$

where Γ_1 denotes the useful single pass gain of the excitation laser in excess of the nonuseful components such as scattering and absorption losses. For power extraction by means of a transmitting mirror, of reflectance R, Γ_1 represents the single pass loss associated with the mirror or $\Gamma_1 = -1/(2d \ln R)$. For steady-state operation the single pass loss and gain are equal.

For the optimum excitation of the intracavity laser

$$\Gamma_1 = N_0 \sigma_p d$$

from which

$$P_p(0) = 2 N_0 \sigma_p d P_{1C}^+(0).$$

Substituting this expression into (2.12) yields

$$G_0 = \frac{4\gamma\tau P_p(0)}{b_p hc},$$

which is identical to the maximum gain obtainable from the externally excited dye laser given by (2.11). On the basis of these results, obtained by PIKE [1971], there appears to be no advantage in gain to be realized by placing the dye laser inside the exciting laser cavity. The intracavity geometry may offer some other advantages, however. One advantage with respect to the externally excited device is the uniform dissipation of heat produced by excitation over the active volume. This tends to alleviate hot spots, a serious problem in high-power cw dye lasers.

The on-axis gain expression derived above is valid when the diameter of the laser beam is much smaller than the diameter of the excitation beam. When the beam diameters become comparable it is necessary to integrate the intensity of the laser beam over the whole wave front, that is over all r, to determine the gain. The calculation yields a result which is somewhat less than G_0 since the active medium is not so highly excited for $r > 0$ as for $r = 0$. The off-axis gain is lower than the gain on the $r = 0$ axis.

A convenient way of expressing the relative sizes of the excitation and laser beams is to take the ratio of their areas at the entrance face of the active medium,

$$\varrho = \frac{\pi w_{p0}^2}{\pi w_{L0}^2} = \frac{\lambda_p b_p}{\lambda_L b_L}.$$

Pike (1971) has shown that the net single pass gain depends upon ϱ as

$$G(\varrho, G_0) \cong \frac{G_0}{1 + \varrho} \tag{2.13}$$

to a good approximation. This approximation is derived for the case in which the excitation and laser beam radii do not change greatly over the active region, a situation which holds in most cases of practical interest.

With reference to (2.13) it should be pointed out that $G_0 \propto P_p(0)$, the incident excitation power even for $\varrho \neq 1$. The gain required to reach laser threshold is fixed by laser system parameters. Thus, (2.13) expresses the dependence of the pump power required to reach threshold upon the mode-match parameter ϱ. Note that the threshold pump power is a minimum when the laser beam is much smaller than the pump beam. The discussion that follows will show that the value of ϱ giving minimum threshold is not consistent with maximum efficiency. In fact, the efficiency of the low threshold device is very low since the excitation beam is used very inefficiently.

Pike [1971] has also developed expressions for the dependence of laser output power and efficiency upon the mode matching parameter. In the derivation it is assumed that saturation of the ground state absorption could be neglected, that the excited state population follows a gaussian distribution across the active region, and that the active region is short in comparison with the pump and laser beam confocal parameters. These are good assumptions for a practical cw laser system.

Under these conditions the laser output power is given by

$$P_L = \frac{\Gamma_1}{\Gamma} \left(1 - e^{-N_0 \sigma_p d}\right) \frac{\varrho}{\sqrt{1 + \varrho^2}} (P_p - P_t) \tag{2.14}$$

where Γ is the total single pass loss at the laser wavelength including useful and nonuseful components, such as scattering, singlet and triplet state absorption, extraneous reflections and so forth. Γ_1 is the useful loss associated with mirror transmission, as discussed previously, and

$$P_t = \frac{\Gamma(1 + \varrho) b_p hc}{4 \sigma_L \tau [1 - \exp(-N_0 \sigma_p d)]} \quad \text{is the threshold power.}$$

The slope efficiency is found from (2.13) to be

$$\eta_{\text{slope}} = \frac{\Gamma_1}{\Gamma} \left(1 - e^{-N_0 \sigma_p d}\right) \frac{\varrho}{\sqrt{1 + \varrho^2}}.$$

To illustrate the manner in which the slope efficiency and threshold power change with ϱ (2.14) is plotted in Fig. 2.8 for a dye laser having a set of parameters corresponding to a practical, low power, cw dye laser.

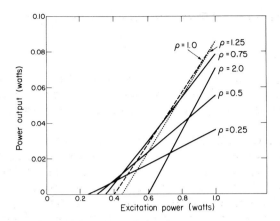

Fig. 2.8. Calculated dependence of dye-laser output upon excitation power. The curves illustrate the dependence of laser threshold and slope efficiency upon the choice of ϱ, the ratio of the areas of the laser and excitation beams at the beam waists (PIKE, 1971)

Figure 2.8 illustrates graphically the trade-off between threshold and operating efficiency. Note that the slope efficiency becomes constant for large values of ϱ. The operating efficiency of the laser, defined as $\eta_{0p} = P_L/P_p$, is a maximum for some value of ϱ which depends upon the available input power. It appears from the diagram that there is never an advantage to be gained from having $\varrho \gtrsim 1.25$, except perhaps for very high power devices. For low pump power, small ϱ is favored, while for an input power of several watts the optimum value of ϱ appears to be about 1.25.

The results displayed in Fig. 2.8 also give some indication of the dilemma faced by the designer wishing to maximize efficiency and tuning range of a tunable laser. For maximum tunable range, P_t should be minimized. However, this results in an inefficient design. The desirability of maximum tunable range and maximum efficiency must be traded off against one another.

2.2. Tuning of the cw Dye Laser

The gain analysis of Section 2.1 centered upon a laser resonator geometry of the form shown in Fig. 2.3. It was assumed that the mirror spacing of the resonator could be adjusted to give the proper mode spot size in the active region. These assumptions are valid but a two-element resonator such as that of Fig. 2.3 possesses some practical disadvantages for tunable systems as TUCCIO and STROME (1972) have pointed out. These are best illustrated by an example.

For a nearly hemispherical cavity, such as that shown in Fig. 2.3, the radius of the TEM_{00} mode at the flat mirror will be given approximately by (BOYD and GORDON, 1961)

$$w_0 = \left(\frac{\lambda}{\pi}\right)^{1/2} [D(X_2 - D)]^{1/4}$$

if $w_0 \gg \lambda$. The sensitivity of w_0 to changes in D is displayed by differentiating w_0 with respect to D to obtain

$$\frac{dw_0}{dD} = \left(\frac{\lambda}{\pi}\right)^2 \left(\frac{X_2 - 2D}{4w_0^3}\right).$$

To a good approximation, therefore

$$\frac{\Delta w_0}{w_0} = \left(\frac{\lambda}{\pi}\right)^2 \left(\frac{X_2 - 2D}{4w_0^4}\right) \Delta D.$$

The desired spot size is fixed by the available excitation power, as discussed previously. For $w_0 \approx 5\,\mu m$ with a laser operating at 600 nm, the change in mirror spacing which will double the spot size may be easily calculated. For the minimum spot size, $D \approx X_2$ so that a doubling of w_0 occurs for

$$\Delta D = -\frac{3.4 \times 10^{-4}}{D}\,cm^2.$$

If D is 0.3 cm, $\Delta D \approx 10\,\mu m$. The tolerance on the position of the curved mirror will actually be somewhat less than this value. Although this tolerance is manageable, the cavity length of 0.3 cm does not allow space for the insertion of an intracavity tuning element. For this purpose D must be extended to about 15 cm. For a cavity of this length ΔD drops to approximately 0.15 μm, an unreasonable tolerance on the location of the curved mirror. From these considerations it is seen that a two-element resonator is not suitable for tunable cw dye lasers.

The dimensional tolerances can be relaxed by the use of a three-element resonator. Two types of three-element resonators have been studied in some detail. The first of these systems, using transmitting optics, has been described by TUCCIO and STROME (1972) and by HERCHER and PIKE (1971). A system using reflecting optics has been designed by KOHN et al. (1971) and has been discussed by KOGELNIK et al. (1972). The design considerations for these two types of systems will be discussed in turn.

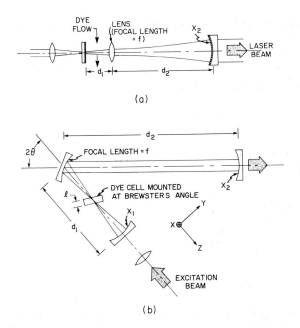

Fig. 2.9. a Three-element dye-laser resonator using transmitting optics. b Three-element dye-laser resonator using reflecting optics (KOHN et al., 1971)

The basic form of the three-element resonator using transmitting optics is shown in Fig. 2.9a. In this system the curved mirror of the two-element resonator is replaced by a lens plus a mirror with long radius of curvature. TUCCIO and STROME (1972) have shown that with this arrangement the position of the long-radius mirror is not nearly as critical as with the two-element resonator. The position of the lens is critical, however, though it may be rigidly fixed with respect to the first, flat mirror when the laser is constructed and does not need subsequent adjustment.

TUCCIO and STROME [1972], following the treatment of KOGELNIK [1965] utilized the complex beam parameter q to describe the propagation of the Gaussian TEM_{00} mode within the laser resonator. The beam parameter is defined by the relation

$$\frac{1}{q} = \frac{1}{X} - \frac{i\lambda}{\pi w^2}$$

where X is the radius of curvature of the wavefront at any point along the beam, and w is the radius of the mode at the same point.

The minimum radius of the beam, w_0, occurs at the flat mirror. The complex beam parameter q_0 at this surface is given by

$$q_0 = i\pi w_0^2/\lambda$$

since X is infinite. Utilizing q_0 the beam can be traced back through the system and, from a relationship given by KOGELNIK (1972), the complex beam parameter can be found at any point as a function of f and the spacing between the components, namely

$$\frac{1}{q} = \frac{-q_0/f + (1 - d_1/f)}{(1 - d_2/f)q_0 + (d_1 + d_2 - d_1 d_2/f)}. \tag{2.15}$$

TUCCIO and STROME demonstrated that the spot size at the flat mirror is stationary with respect to changes in d_1 and d_2 when $d_1 = f$, if the curvature of the mirror, X_2, is assumed to be the curvature of the wavefront and $d_2 = X_2/2 + f$. The sensitivity of the spot size to changes in d_1 and d_2 for several values of X_2, as obtained from (2.15) is shown in Figs. 2.10 and 2.11, respectively.

From Fig. 2.10 it is apparent that the allowable tolerance in the placement of the curved mirror increases as the radius of curvature of the mirror increases. The advantage with respect to the two-element cavity, in which the tolerance decreased with increasing mirror radius, is clear.

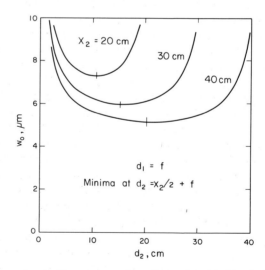

Fig. 2.10. Dependence of laser beam waist radius or "spot size" at the flat mirror of the resonator of Fig. 9a upon the spacing between the collimating lens and the curved mirror. The radius of curvature of the mirror, x_2, has been taken as a parameter. The curves are computed from (2.15) using a value of 0.53 cm for f (TUCCIO and STROME, 1972)

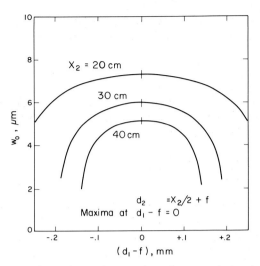

Fig. 2.11. Dependence of the laser beam waist radius upon the spacing between the flat mirror and the collimating lens. The radius of curvature of the mirror at the end of the long arm of the cavity is taken as a parameter. The curves are computed from (2.15) using a value of 0.53 cm for f (TUCCIO and STROME, 1972)

From Fig. 2.11 an evaluation of the tolerance on the placement of the lens can be obtained. It appears that the longer cavity is again the less critical, though the location of the lens must be maintained within approximately ± 0.1 mm of the position for minimum spot size. This value is very much less critical even than the tolerance on placement of the curved mirror in a two-element resonator several millimeters long.

In the preceding brief analysis it has been assumed that the intra-cavity lens is a thin lens. For a practical system the thickness of the lens will be comparable to the short focal length required. Thus, it will be necessary to use a thick-lens analysis. The modifications of the analysis resulting from this change were discussed by TUCCIO and STROME [1972].

The principal disadvantage associated with the resonator of Fig. 2.9a is that the lens introduces two extra reflecting surfaces, with their attendant loss, into the cavity. This loss may be minimized by antireflection coating of the surfaces. However, even small reflections may be serious for some purposes, such as mode-locking (see Chapter 3). In these cases a three-element resonator using reflecting optics is advantageous.

A three-element reflecting cavity is shown schematically in Fig. 9b. The analysis of this system, as given by KOGELNIK et al. (1972), is similar in many respects to that for the three-element cavity with transmitting optics. The principal difference in the analysis results from the folding of the system, which requires that the center mirror is operated off-axis. This introduces astigmatism. The rays in the plane of the drawing are focused at a different point than those perpendicular to the plane of the paper. The Brewster-angle dye cell also introduces astigmatism. KOGELNIK et al. (1972) have shown that the thickness of the dye cell and the folding angle θ of the cavity can be adjusted so that the astigmatism of the mirror and dye cell cancel. Furthermore, the parameters of the compensated cavity may be such that the focal spot size is small and has a position which is relatively insensitive to the length of the short leg of the cavity. The treatment and results developed by KOGELNIK et al. [1972] will only be outlined here. The reader is referred to the original paper for details of the analysis.

The three-element cavity will be stable for some range of variation about the spherical cavity separation $d_{1s} = X_1 + f$. The deviation of d_1 from this value is taken as δ where

$$d_1 = X_1 + f + \delta.$$

The limits upon δ for stability of the cavity are taken as δ_{max} and δ_{min} so that

$$d_{1\,max} = X_1 + f + \delta_{max}$$
$$d_{1\,min} = X_1 + f + \delta_{min}$$

with $\delta_{max} = f^2/(d_2 - f)$ and $\delta_{min} = f^2/(d_2 - R_2 - f)$. These stability limits are derived from the consideration of the equivalent, two-element, spherical resonator. Note that δ_{min} is negative or zero, since $X_2 \geq d_2 - R_2 - f$. KOGELNIK et al. (1972) show that the focal spot varies as d_1 is changed throughout its stability range approximately in accordance with

$$\left(\frac{\pi w_0^2}{\lambda}\right)^2 \approx (\delta_{max} - \delta)(\delta - \delta_{min}). \tag{2.16}$$

The spot radius w_0 vanishes at the limits of the stability region and has its maximum size near the center of the range. As a result of astigmatism the focal length of the center mirror is different for rays in the xz (sagittal) plane from that for rays in the yz (tangential) plane. For rays in the xz plane $f_x = f/\cos\theta$ and for the yz plane $f_y = f \cdot \cos\theta$, where f denotes the focal length of the mirror. The focal-spot size is dependent upon position and upon focal length.

Neglecting, for the moment, the presence of the dye cell, the spot size as a function of position in the sagittal and tangential planes is shown for one set of cavity parameters in Fig. 2.12a. The astigmatism resulting from $f_x \neq f_y$ is clearly seen. Such an optical cavity is unstable

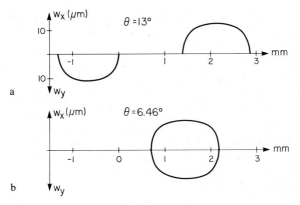

Fig. 2.12. a Focal-spot size and position for both sagittal (xz) and tangential (yz) planes of the resonator in Fig. 2.9b. The refractive index of the dye cell has been neglected in plotting the spot sizes according to (2.16). b Compensation of the astigmatism of Fig. 2.12a by the Brewster-angle-mounted dye cell. The dye cell has been taken to have a thickness $l = 0.15$ cm and refractive index $n = 1.45$. (In both plots the values $x_1 = 5$ cm, $x_2 = \infty$, $l = 5$ cm, and $d_2 = 190$ cm have been used (KOGELNIK et al., 1972)

since the sagittal and tangential rays focus at different points. There is no value of d_1 for which both rays form stable modes.

The Brewster-angle cell introduces a compensating astigmatism which may be used to stabilize the resonator. The optical path length for the xz and yz rays is different in the Brewster-angle cell. The effective thickness for xz plane rays is given by

$$d_x = \frac{t}{n^2} \sqrt{n^2 + 1}$$

and that for yz plane rays by

$$d_y = \frac{t}{n^4} \sqrt{n^2 + 1} ,$$

where n is the refractive index, and t is the actual thickness of the cell. KOGELNIK et al. (1972) demonstrated that the choice of the angle θ and cell thickness t such that

$$Z K t = \frac{f}{2} \sin \theta \tan \theta ,$$

where $K = [(n^2 - 1)/n^4] \sqrt{n^2 + 1}$, results in a complete overlap of the stability regions. The waist radii for the saggital and tangential rays of a cavity compensated in this manner are plotted in Fig. 2.12b for a cavity having the same parameters as that of Fig. 2.12a.

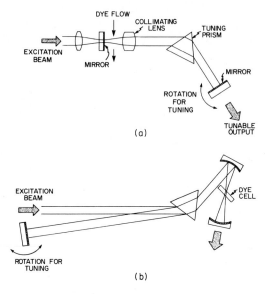

Fig. 2.13. Modification of the three-element resonators of Figs. 2.9a and 2.9b to include a tuning element (prism) in the long arm of the resonator

In some cw dye-laser systems a spherical resonator has been used to separate the dye cell from the optical components of the resonator. In this system the flat mirror of Fig. 2.9a is replaced with a mirror having a short focal length so that the beam waist falls within d_1 rather than at the mirror surface. This arrangement introduces extra reflections, associated with the dye cell windows, into the cavity. It is generally not possible to eliminate these by mounting the cell at Brewster's angle because of the astigmatism of the dye cell, as discussed above. In the cavity using transmitting optics there is no simple way to compensate for the astigmatism. Modification of the three-element resonators to provide for tuning are shown in Figs. 2.13a and 2.13b. The space within the cavity between the collimating element and the resonator mirror is available for insertion of a tuning element. A prism is usually used.

The input-mirror requirements for the cavity with reflecting optics shown in Fig. 2.13b are less stringent than for the cavity with transmitting optics. The input mirror needs only to have high reflectance, while in the arrangement of Fig. 2.13a the input mirror must be highly transmitting at the excitation wavelength. This is avoided in the arrangement shown in Fig. 2.13b by dispersing the excitation beam with the tuning prism (Dienes et al. 1972). The disadvantage of this system is that only a single output line of the exciting laser may be used. Thus the total power obtain-

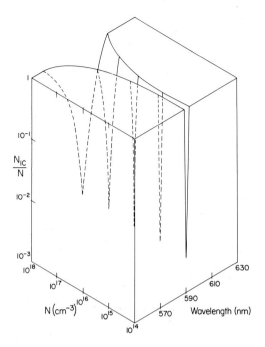

Fig. 2.14. Extrinsic critical inversion surface modified by the inclusion of a dispersive element within the resonator. A mirror with a reflectance spectrum $R = \exp[-(\lambda - \lambda_0)^2/\Delta\lambda^2]$ has been used, where $\Delta\lambda = 5$ nm and $\lambda_0 = 590$ nm

able from a multiline source is not available. This is not a problem with the cavity design of Fig. 2.13a.

The effect of system parameters upon the laser with a dispersive resonator can be visualized by reference to the critical inversion surface. The surface for a laser with an active length of $d = 0.2$ cm, and a dispersive mirror having a reflectance $R_2 = R_0 \exp[-(\lambda - \lambda_0)^2/\Delta\lambda^2]$ where $R_0 = 1$, $\lambda_0 = 590$ nm and $\Delta\lambda = 5$ nm, is shown in Fig. 2.14. In this figure, one of the axes is dye concentration. It is seen that the width of the valley narrows as the dye concentration decreases. The bottom of the valley intersects the critical inversion surface with nondispersive mirrors along the $\lambda = 590$ plane. The plot suggests that the greater laser output line stability is associated with a low concentration of dye. This result is understood with reference to (2.7). As N decreases the $r(\lambda)/N$ term becomes dominant with respect to the $\sigma_s(\lambda)$ term.

The $1/e$ linewidth, 10 nm, used for the tuning element in Fig. 2.13 is considerably broader than that associated with a practical tuning element, which should have a linewidth of less than 1 nm. The valley of Fig. 2.14

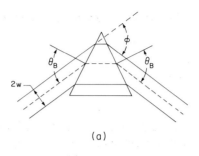

(a)

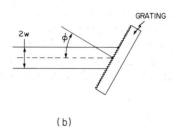

(b)

Fig. 2.15. Detail of the laser beam intersecting the tuning element of a tuned laser. These diagrams define the geometrical parameters used in estimating the linewidth of a laser tuned by a (a) prism or (b) grating

would be correspondingly narrower than that shown. This relatively large value was chosen only for illustrative purposes.

The linewidth limit of the cw dye laser will be determined in part by the linewidth of the tuning element and in part by statistical processes. Each of these contributions will now be considered.

The dispersive portion of a resonator tuned by a single prism is shown in Fig. 2.15a. For this arrangement, the linewidth $\Delta\lambda$ will be given by

$$\Delta\lambda = \frac{d\lambda}{dn} \cdot \frac{dn}{d\varphi} \Delta\varphi = \frac{1}{\left(\frac{dn}{d\lambda}\right)\left(\frac{d\varphi}{dn}\right)} \Delta\varphi,$$

where n denotes the refractive index of the prism, and φ is defined by the figure.

If the laser is operating in its lowest-order (TEM$_{00}$) transverse mode, the divergence of the beam within the cavity, $\delta\varphi$ will be approximately $\lambda/(2w)$, where w is the radius of the beam at the collimating element of the cavity. It was observed (HERCHER and PIKE, 1971) that the laser ceases to oscillate when the mirror misalignment is about $\delta\varphi/20$. This figure may be taken as the angular deviation of the resonating beam from the resonance angle φ. The oscillating linewidth, therefore, will be approximately

$$\Delta \lambda = \frac{1}{\left(\dfrac{dn}{d\lambda}\right)\left(\dfrac{d\varphi}{dn}\right)} \frac{\lambda}{40w}.$$

Using the values for a phosphate flint prism (TUCCIO and STROME, 1972) ($n \cong 1.98$ at 589 nm) cut at BREWSTER'S angle and operated at minimum deviation, with $2w = 0.2$ cm, the expected linewidth is about 0.1 nm. The linewidth may be narrowed further by incorporating etalons in the cavity in addition to the prism.

A similar argument may be applied to a system tuned with a grating. The grating is assumed to be mounted in the Littrow mounting, as shown in Fig. 2.15b. In this case the linewidth will be given by

$$\Delta \lambda = \frac{\lambda}{\tan \varphi} \Delta \varphi = \frac{\lambda^2}{40w \tan \varphi}.$$

If $\varphi \approx 30°$, and $\lambda = 600$ nm, the expected linewidth is about 0.016 nm. It should be noted that the linewidth is independent of the ruling spacing of the grating. It depends only on the angle at which the grating is used. A coarse grating used at high order is completely equivalent to a fine grating used at low order. There may be an advantage in the use of the coarse grating in that it is not so susceptible to damage as the fine grating. The linewidth associated with the grating is seen to be much less than that associated with a single prism. For the cw laser, however, the prism is to be preferred since the losses are lower than those of a grating. Efficient pulsed dye laser systems have been made by using a grating in conjunction with a dielectric mirror (BJORKHOLM et al., 1971). This technique should be applicable to cw lasers as well.

Of the statistical effects which influence the linewidth, three have been considered (HERCHER and SNAVELY, 1972): Spontaneous emission or phase diffusion, Brownian motion of the resonator, and statistical fluctuations in the density of the active medium. Only the last of these is large enough to be important in practical systems.

The density (fluctuations) in a liquid due to the statistical fluctuation in the number of molecules per unit volume have been treated by many authors. An early treatment by EINSTEIN (1910) yields the rms fluctuation in density, ϱ, as

$$\left(\frac{\Delta \varrho}{\varrho}\right) \approx \sqrt{\frac{kT\beta}{V_c}},$$

where β is the isothermal compressibility, T is the absolute temperature, k is Boltzmann's constant, and V_c is the volume under consideration. The appropriate volume for the dye laser is the active volume of the system, as shown in Fig. 2.6. In determining the frequency shift associated with these density changes it is assumed that the polarizability is proportional to the number of molecules per unit volume. The refractive index is proportional to the square root of the polarizability so that

$$\left(\frac{\Delta n}{n}\right) = \frac{1}{2}\left(\frac{\Delta \varrho}{\varrho}\right).$$

Putting values appropriate to a water solution into this equation, it is found that $\Delta n/n \approx 2.3 \times 10^{-9}$ if a value of $10^{-7}\,\mathrm{cm}^3$ is chosen for V_c. To determine the frequency change Δv produced by this fluctuation in refractive index, with reference to Fig. 2.3, it is noted that for an active length of 0.2 cm and a total length of 30 cm,

$$\frac{\Delta v}{v} = \frac{\Delta d}{D} = \frac{\Delta n d}{n D} \cong \frac{2.3 \times 10^{-9} \times 0.2}{300} = 1.53 \times 10^{-12}.$$

For a laser operating in the yellow at $5 \times 10^{14}\,\mathrm{Hz}$, $\Delta v \approx 800\,\mathrm{Hz}$. This value is much smaller than the linewidth observed to date with any practical system. Reasons for this discrepancy will be discussed in the following section.

2.3. Performance of Experimental Systems

Reports on the operation of cw dye lasers have been published by several laboratories. Since the initial demonstration of cw operation (PETERSON et al., 1970), power output, tuning range, linewidth, stability, and efficiency have improved steadily. The state of the art, as of early 1973, is summarized in Table 2.1 Most of the capabilities listed can be expected to improve further with continued research.

The power output is limited at present by the power available from the gas lasers used for excitation. As more powerful lasers are developed, the output of the cw dye laser will increase.

The tuning range of the cw dye laser has recently been extended toward the blue region of the spectrum by TUCCIO and DREXHAGE (1973). The argon lines at 351 nm and 364 nm were used to excite coumarin derivative dyes. Operation at the short wavelength end of the range, 420 nm, is restricted to long pulses by the power available in these lines. Power output of the cw dye laser will certainly be increased and extended in the blue and near uv spectral regions. Extension of the output wavelength beyond 710 nm requires more powerful red and near infrared sources. It is possible that light emitting diodes will be useful for exciting dye lasers in this spectral region.

TUCCIO and STROME (1972) have described the operation of a three-element dye laser. The device had an optical arrangement very similar to that of Fig. 2.11a. In this system a phosphate flint Brewster-angle prism was used for tuning. The device was excited with the 4-watt output of an argon laser at 514.5 nm. The laser mode and pump waist diameters were matched at 12 μm. The active length of the system was 0.3 cm.

The output power as a function of wavelength is shown for several dyes in Fig. 2.16. The tuning curves of Fig. 2.16 were obtained with water solutions of the dyes rhodamine 6G and rhodamine B using two different additives. The refractive index of water depends much less strongly upon temperature than that of methanol or ethanol. The use of water solutions of dyes tends to reduce the laser instability caused by nonuniform heating of the active region.

Many of the best laser dyes do not dissolve readily in water (see Chapter 4). Rhodamine 6G, for instance, tends to dimerize in water. To prevent this it is necessary to add a deaggregating agent to the dye solution. In Fig. 2.16, curve 1, hexafluoroisopropanol was used as the deaggregating agent. For curves 2 and 3 the surfactant Ammonyx LO was used. It was found that these additives also tend to quench the triplet

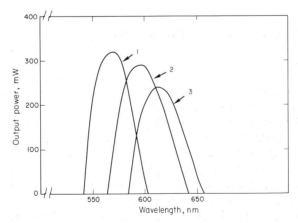

Fig. 2.16. Power output vs wavelength for three different dye solutions used in a tunable system having the geometry of Fig. 2.13a. Curve 1 was obtained with a 2×10^{-4} M solution of rhodamine 6G in 3:1 water-hexafluoroisopropanol; curve 2 represents a 3×10^{-4} M solution of rhodamine 6G in water plus 5% Ammonyx LO, and curve 3 was the output from a 3×10^{-4} M solution of rhodamine B in 3:1 water-hexafluoroisopropanol. The excitation of the dye laser was 4 watts at 514.5 nm (TUCCIO and STROME, 1972)

state of the dye molecules. The reasons for this behavior are not yet understood.

Interestingly, one dye may produce quite different fluorescence spectra, as reflected by the range of the tuning curve, in different solvents. In Fig. 2.16, curves 1 and 2 are both rhodamine 6G in water. The different tuning curves result from the use of different deaggregating agents.

The total tuning range of about 70 nm for a given dye is typical of the performance of the cw dye laser. The efficiency of this laser however has not been optimized in this design, as pointed out by TUCCIO and STROME (1972). The efficiency and power output of the system were found to be degraded by birefringence in the optical components of the dye cell. If the effect is not eliminated, spurious reflection losses will occur at the tuning prism since the polarization of the laser beam will not be determined completely by the Brewster-angle surfaces of the prism. This effect is very likely to occur when sapphire elements are used in the dye cell. Modification of the system to eliminate the effects of birefringence in the dye cell has yielded a slope efficiency of about 45% and over-all efficiency of approximately 30%. Similar results have been obtained by DIENES et al. (1972) with three-element reflecting cavities.

TUCCIO and STROME (1972) developed an analysis similar to that of PIKE (1971), which was reviewed in Section 2.1, and have compared the measured and calculated performance characteristics for tuned and untuned cw dye lasers. Figure 2.17 shows the comparison between calculated and observed power output vs power input. The output with

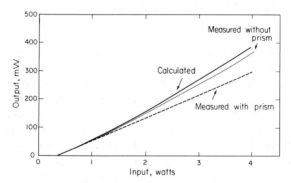

Fig. 2.17. Comparison of measured and calculated output power as a function of input power for a cw dye laser using the geometry of Fig. 2.13a. The data of Fig. 2.2 were used in the analysis with the following additional dye and resonator parameters: $N = 1.8 \times 10^{17} \, \text{cm}^{-3}$, $W_0 = W_{p0} = 6\mu\text{m}$, $d = 0.3 \, \text{cm}$, $R_1 = 0.992$, $R_2 = 910$, output mirror transmittance $= 0.03$. A single-wavelength excitation at 514.5 nm was assumed. Absorption loss of the intracavity lens was included in the analysis, though the absorption due to the prism was not. The slight lowering of efficiency in the prism-tuned system is due to the loss of the prism (TUCCIO and STROME, 1972)

the tuning prism in the cavity is seen to be slightly lower than that without the prism owing to the losses associated with the prism. The difference between the calculated and measured results is attributed to scattering losses which have not been included in the calculations. Nevertheless, the agreement is seen to be fairly good. Figure 2.17, and similar results obtained by PIKE (1971), strongly support the validity of the rate-equation approach to the analysis of the cw dye laser.

Measurements were made with the same system to determine the effect of dye flow velocity upon the output characteristics. It was found that the amplitude noise decreases and the power output increases markedly as the flow rate increases from 2 m/sec to 12 m/sec. Experiment has also shown that the laser threshold decreases with increasing flow velocity, presumably because of the physical removal of triplet-state molecules from the active region. With transit times of less than 10^{-6} sec across the active diameter, corresponding to a flow velocity of about 10 m/sec, "mechanical quenching" becomes a useful technique for control of the triplet-state population. Flow velocities of up to 100 m/sec do not seem unreasonable, in which case the transit time across the 10 μm active diameter would be about 10^{-7} sec. This is short enough to be used as the primary means for control of the triplet state population and permits consideration of dyes for which no chemical quenchers are known. With such large flow velocities, however, the flow geometry of the system must be carefully designed to avoid cavitation in the dye cell. In addition to possible mechanical damage to the system, cavitation would produce severe degradation of the optical quality of the resonator.

Several authors have reported on the single longitudinal mode operation of cw dye lasers (HERCHER and PIKE, 1971a; SCHRÖDER et al., 1972). In all cases an arrangement similar to that shown in Fig. 2.13a or b has been used with an etalon mounted in the long arm of the cavity in addition to the prism. In the system reported by HERCHER and PIKE a 2-mm uncoated glass etalon was used in conjunction with a single prism. The linewidth of the system without the etalon was approximately that given by geometrical considerations, as discussed in the preceding section. With the etalon in the cavity the short term (several seconds) linewidth was reduced to about 35 MHz. Figure 2.18a shows a scanned Fabry-Perot interferometer trace of the dye laser output. The scanned output of a multimode He-Ne laser is shown in Fig. 2.18b for comparison. HERCHER and PIKE (1971) observed that the line was stable to within 180 MHz over a period of minutes. Using a similar system, SCHRÖDER et al. (1972) have reported a short-term (one second) stability and linewidth of less than 10 MHz.

In more recent experiments HARTIG and WALTHER (1973) have reported a somewhat narrower linewidth and greater stability. In their

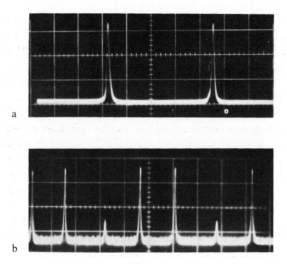

a

b

Fig. 2.18. a Single longitudinal mode output of a cw dye laser, as displayed by a scanning Fabry-Perot interferometer. The single mode output was obtained by use of an intracavity etalon (2 mm glass, uncoated) in addition to the prism in an arrangement similar to that shown in Fig. 2.13a. The free spectral range of the interferometer, as displayed, is 1.5 GHz. The laser linewidth is 35 MHz. b Interferometer trace of the multimode output of a He–Ne laser for comparison with (a) (Hercher and Pike, 1971a)

experiments a spherical three-element cavity with transmitting optics similar to that described by Hercher and Pike (1971) was used. A linewidth of less than 20 MHz was deduced from measurement of the Na resonance line hyperfine structure. The laser output was locked to the Na D_1 line by conventional techniques in one experiment. Under these conditions a long term drift of the laser line center of less than ± 1.5 MHz, with respect to the sodium line, was achieved. The laser produced a single mode output of 10 mW with an excitation power of 1.2 watts at 514.5 nm.

The linewidths mentioned in the preceding paragraphs are considerably greater than the linewidth expected on the basis of statistical processes. In the preceding section arguments were presented for an ultimate linewidth of less than 1 kHz. The discrepancy does not seem to be due to the lack of dispersion in the resonator, but rather to problems associated with the flow of the dye solution through the active region. The dye cells utilized to date have not been designed to minimize cavitation or turbulence within the active region. The large observed linewidth suggests that there are gross instabilities in the dye stream as it flows through the active region.

Variation in the excitation power may also cause an artificially broad dye laser line. Most commercial argon ion lasers, of the sort used to excite dye lasers, have amplitude noise of several percent. This noise will produce temperature fluctuation in the active region which, in turn, may cause frequency fluctuations in the dye laser output. Narrow laser linewidth would seem to require stabilization of the excitation laser.

The excitation brightness necessary to excite a cw dye laser has led to operational problems in some cases. With an excitation power of several watts, damage may occur at the surface of the dye cell window at the beam waist. The severity of this problem seems to increase with dye solutions containing triplet quenching and dye dispersing additives. Cell windows also seem to be more susceptible to damage when impure or unclean solvents and dyes are used.

One approach to a solution to the window burning problem has been the use of a windowless dye "cell" in the form of a free-flowing stream of dye from a nozzle. This technique was first described by RUNGE (1972) and has since been further developed (RUNGE and ROSENBERG, 1972).

Use of this technique places some restriction upon the choice of solvents and dyes. Unfortunately, the most favorable dye solvents, such as water and alcohols, do not seem to form stable streams from simple nozzles. To date experiments have been performed with ethylene glycol and glycerol-based dye solutions. Although the thermal properties of these solvents are undesirable, good performance is claimed for cw dye lasers using the free stream.

2.4. Conclusion

In this chapter the factors affecting power output, efficiency tuning range, stability and linewidth of cw dye lasers have been discussed. Performance analysis has been approached from the rate-equation point of view. This approach has been found to give an acceptable description of the cw dye laser. At present it seems that no serious

Table 2.1. Summary of cw dye-laser characteristics

Power output	2 W
Tuning range	420–710 nm
Linewidth	$\approx$ 2 MHz (tuned)
	$\approx$ 3 nm (untuned)
Efficiency	$0.35 \dfrac{\text{optical power out}}{\text{excitation power in}}$
Mode-locked pulse-width	$\approx$ 1.5×10^{-12} sec

attempts to apply the more powerful semiclassical or quantum electro-dynamic analyses to the dye laser have been made. Progress in this theoretical direction is to be anticipated.

Commercial cw dye lasers with specifications comparable to those of Table 2.1 are becoming available. Powerful ultraviolet lasers as excitation sources for cw dye lasers operating in the blue spectral region are just becoming available at this time. As other new excitation sources become available extension of the tuning range into the near infrared is to be expected.

3. Mode-Locking of Dye Lasers

C. V. SHANK and E. P. IPPEN

With 11 Figures

One of the most attractive features of organic dye lasers is their ability to produce ultra-short optical pulses. Since the frequency bandwidth required to produce a pulse of picosecond duration is still much less than that of a typical dye emission band, a dye laser can be a source of wavelength-tunable picosecond pulses. Dye lasers are also the only devices at present capable of producing such pulses on a continuous, highly repetitive basis.

The method by which ultra-short pulses are generated in a laser is called mode-locking. Mode-locking is not new to dye lasers and has been studied extensively in a variety of other laser systems (SMITH, 1970). In the first part of this chapter we will review the basic principles of mode-locking and include a brief description of some pulse measuring techniques. In subsequent sections we will describe experimental results obtained with both high-power, pulsed dye lasers and the recently developed cw laser. Finally, we will discuss some of the unresolved questions regarding the way picosecond pulses are produced by dye lasers and compare dye lasers in this respect with other systems.

3.1. General Background

A typical laser consists of an optical resonator formed by mirrors and a laser gain medium inside the resonator. Although the gain medium determines approximately the wavelength of laser operation, properties of the resonator define more precisely the laser frequencies (KOGELNIK and LI, 1966). Common curved mirror resonators support a variety of laser modes some of which may have different field distributions normal to the resonator axis (transverse modes). Each of these transverse modes has an infinite set of eigenfrequencies (longitudinal modes) separated in frequency by an amount $c/2L$ where L is the optical length of the resonator and c is the speed of light. Generally it is possible to aperture the resonator in such a way as to discriminate against all transverse modes except the lowest order which has a simple Gaussian profile.

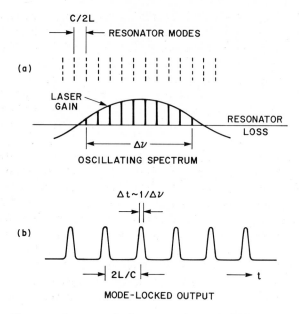

Fig. 3.1. a The spectral content of a laser operating on a single transverse mode and a number of different longitudinal modes. b The temporal output of the laser in (a) with all modes locked in proper phase

Unless some form of highly selective frequency discrimination or an extremely short resonator is used, this Gaussian profile laser output still consists of a number of regularly spaced ($c/2L$) frequencies. The number which oscillates is limited by the bandwidth Δv over which the laser gain exceeds the loss of the resonator. A typical laser output spectrum is illustrated in Fig. 3.1a.

Since the output of a laser consists of a number of frequency components, it is clear that the amplitude of the output may vary in time in a variety of ways depending on the relative phases and amplitude of these frequency components. If nothing fixes these parameters, random fluctuations cause the output to vary with time even though the average power may remain relatively constant. On the other hand, if the modes are somehow forced to maintain a fixed phase and amplitude relationship, the output will be a well-defined function of time and the laser is said to be "mode-locked". It is possible, for example, to fix the mode amplitudes and phases in such a way that they produce a constant amplitude, frequency modulated (FM) output signal. More importantly, for our discussion, mode-locking can produce a strongly amplitude modulated output consisting of a regularly spaced train of pulses. The pulses have a width Δt which is approximately equal to the reciprocal of the total mode-locked band-width Δv and a temporal periodicity $T = 2L/c$. That is, the ratio of pulsewidth to period is approximately equal to the number of locked modes. Such a mode-locked laser output is illustrated in Fig. 3.1b. In the laser resonator this situation corresponds to a single pulse traveling back and forth between the mirrors with output occurring with each round-trip. It is also often possible to produce multiple pulse behavior (more than one pulse in the resonator) by locking modes separated by some multiple of $c/2L$.

Short pulses can be obtained from a mode-locked laser only if the gain bandwidth is relatively broad. The bandwidth of a typical gas laser for example is only on the order of 10^9–10^{10} Hz which constrains output pulses to be longer than 10^{-10} sec (100 psec). The solid state Nd: glass laser has a bandwidth well in excess of 10^{12} Hz and is capable of producing pulses with sub-picosecond structure. Most interesting from this point of view are dye lasers whose emission bands are so broad (10–100 nm $\cong 10^{13}$–10^{14} Hz) that one can obtain picosecond pulses and still have the freedom of wavelength tuning. The range over which pico-second pulses have been obtained from dye lasers covers continuously the spectrum from 560–700 nm.

There are a number of reasons why it is difficult to produce pulses much shorter than a picosecond even if one has enough available band-width. It is difficult to achieve a uniform mode spectrum with proper $c/2L$ spacing over such a broad spectrum in the face of resonator dis-persion and mode competition in the gain medium. A variety of non-linear effects in the different laser elements also tend to broaden the pulse in time and distort its frequency spectrum. If the oscillating mode spectrum is greater than that required for the observed output pulses of a laser, the situation is fairly complicated. There may be a low-level continuous output between pulses, unobserved noise fluctuations within the pulse envelope, or frequency smearing during the time of the pulse. Noise implies a lack of coherence in the pulse and can lead to erroneous estimates of pulse-width based on nonlinear measurements. Such behavior can often be eliminated by very careful measurement of the pulse, restriction of the oscillating bandwidth, and operation at lower powers. The precise Fourier-transform relationship between the band-width Δv and the pulse-width Δt depends upon the actual pulse shape which is not generally available. For example, a transform-limited Gaussian-shaped pulse has a $\Delta t\, \Delta v = 0.441$ while a Lorentzian pulse has a $\Delta t\, \Delta v = 0.6$. According to convention, Δv and Δt are chosen to mean full width at half maximum (FWHM).

We now consider the ways in which laser mode-locking may be achieved experimentally. It is possible for a laser to self-mode-lock through a nonlinear response of the gain medium itself. More often it is necessary to introduce an external driving force or an additional non-linear element to achieve good mode-locking. A good review on the theory of internal modulation has been presented by HARRIS (1966). A qualitative description of how mode-locking is brought about is the following. Modulation internal to the laser at the frequency $c/2L$, either by an external field or by nonlinear response to the beat between two modes, generates frequency sidebands on either side of each oscillating mode. These sidebands coincide approximately in frequency with the adjacent resonator modes and couple these modes together in a well-defined phase and frequency relationship. If short pulses are to be obtained, a relatively strong modulation is required to force a broad band of modes into oscillation. Over the bandwidth required for very short pulses (picosecond duration) dispersion alters the nominal $c/2L$ mode separation. Driving fields are therefore also required to "pull" the resonator modes to the required frequencies. Spatial hole burning (a suppression of adjacent modes) may also present difficulties (DANIEL-MEYER, 1971). With these complexities a detailed analysis of sideband coupling becomes virtually impossible. An alternative approach is to analyze the development in time of a pulse as it travels back and forth in the resonator through the gain medium and the modulation medium. This approach has provided some valuable insight about the formation of pulses in certain systems (KUIZENGA and SIEGMAN, 1970; FLECK, 1968; KRYUKOV and LETOKHOV, 1970). Nevertheless, a completely satisfying description of mode-locking in dye lasers is not yet available. Some mode-locking results peculiar to dye lasers and some possibly relevant theoretical explanations will be discussed in later sections.

Experimentally, there are various ways in which mode-locking can be achieved (SMITH, 1970). Perhaps the most straight-forward of these is the use of a modulator inside the laser resonator. A modulator placed at one end of the laser resonator and driven at the longitudinal mode-spacing frequency $c/2L$ can provide the necessary mode coupling for locking. In the time domain it may be said that pulses occur because they can best pass through the modulator at a time of least loss. Placing the modulator at positions within the resonator other than at the end can also produce mode-locking if the appropriate driving frequency is used, but this is always a situation which allows multiple pulsing. The chief disadvantage of the modulator technique is that it requires extreme stability of the driving frequency with respect to the resonator length. Internal modulation has not produced ultra-short pulses in mode-locked dye lasers.

Instead of adding a source of loss or phase modulation one can also modulate the gain. While this method is not of much use in most laser systems, it can be particularly suited to dye lasers. Since dye lasers are optically pumped and have very broad absorption bands, it is often possible to drive the dye laser with the output of another mode-locked laser. It is then, of course, necessary to adjust the length of the dye laser resonator to be integrally related to the length of the mode-locked pumping laser. The resulting dye laser pulses can be as short or shorter than the pumping pulses. Picosecond pulses over a variety of wavelengths have been obtained in this manner.

The most widely employed technique for inducing short-pulse mode-locking involves the use of a saturable absorber inside a laser resonator. A saturable absorber is a material whose absorption decreases as light intensity is increased. Thus, a short, high-peak-power pulse experiences less loss in the absorber than a longer, less intense pulse of the same energy. The way in which a saturable absorber produces mode-locking is best discussed in the time domain since the resulting modulation is a function of both the pulse shape and pulse energy. One must consider the way in which the absorber saturates relative to the way in which the gain of the laser medium saturates. In a system where the gain is saturated too easily, high powers are discriminated against and the saturable absorber is less operative. It is also important that the absorber recovery time after bleaching be shorter than the round-trip time of a pulse in the resonator. Otherwise the induced depth of modulation is greatly reduced. The time response (bleaching and recovery times) of a saturable absorber can be much faster than that of any electrically driven modulator. Picosecond pulses have been produced with a variety of gain medium-saturable absorber combinations.

The saturable absorbers most successfully used to produce short-pulse mode-locking have been organic dyes in liquid solution. In a laser resonator the saturable absorber is normally placed at one end to reduce multiple pulsing effects. Quite often the physical arrangement consists of a thin layer of saturable absorber solution in direct contact with an end mirror of the resonator. The optical intensities required to saturate absorption in organic dyes are rather high, so that early advances in their use for mode-locking were restricted to high power, pulsed laser systems. More recently their use has been extended to continuously operated (cw) lasers by the design of optical resonators in which the light is very tightly focussed into the absorber.

The production of ultra-short light pulses by laser mode-locking has also created the challenge of finding ways to measure them. Measurements made with photodetectors and oscilloscopes have a time resolution limited to about 100 psec and are much to slow for observing picosecond pulses. Linear measurements of picosecond pulses have been made using ultra-fast streak cameras. Some of these results are des-

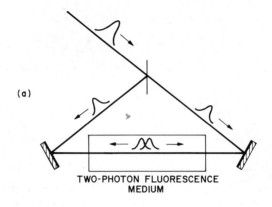

(a)

TWO-PHOTON FLUORESCENCE
MEDIUM

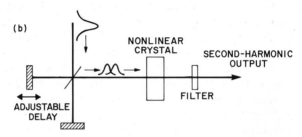

(b)

NONLINEAR
CRYSTAL

SECOND-HARMONIC
OUTPUT

FILTER

ADJUSTABLE
DELAY

Fig. 3.2. Nonlinear correlation techniques for the measurement of ultra-short optical pulses using (a) two-photon fluorescence (TPF) and (b) second-harmonic generation (SHG)

cribed in the next section. The most common techniques for pulse measurement, however, rely on nonlinear optical phenomena which can have time resolutions on the order of 10^{-13} sec. Those used extensively are second-harmonic generation (SHG) (ARMSTRONG, 1967; DIENES et al., 1971a), in a nonlinear crystal and two-photon fluorescence (TPF) (GIORDMAINE et al., 1967), in a material which is transparent to the fundamental laser radiation but can absorb by the two-photon process and then fluoresce. Either of these effects can be used to obtain the autocorrelation of a laser pulse. The way in which this is accomplished in each case is illustrated in Fig. 3.2.

In a TPF measurement the laser pulse is divided into two components which then travel in opposite directions through the TPF medium (normally an organic dye). Fluorescence is observed and generally photographed from the side of the medium. Qualitatively it is understood that, because TPF is a nonlinear process, the most intense fluorescence occurs from the region in the center where the two pulses overlap. It can be shown (KLAUDER et al., 1968) that the fluorescence along the direction of propagation is proportional to

$$f(\tau) = 1 + 2 G^{(2)}(\tau), \tag{3.1}$$

where $G^{(2)}(\tau)$, the autocorrelation function of the pulse intensity, $I(t)$, is given by

$$G^{(2)}(\tau) = \frac{\langle I(t) I(t+\tau) \rangle}{\langle I^2(t) \rangle}. \tag{3.2}$$

The delay τ is related to the distance z along the direction of propagation by $\tau = \dfrac{2z}{v}$, where v is the velocity of light in the medium. Particularly extensive treatments of the way in which $G^{(2)}(\tau)$ depends upon particular experimental parameters have been given by ROWE and LI (1970) and PICARD and SCHWEITZER (1970). For a well-isolated, coherent pulse $G^{(2)}(\tau)$ is 1 at $\tau = 0$ and falls to 0 for large τ. The resulting fluorescence peaks at the point of maximum pulse overlap and exhibits a peak-to-background contrast ratio of 3:1. This theoretical contrast ratio has been verified experimentally (SHAPIRO and DUGUAY, 1969), but is quite often difficult to obtain. If, for example, the pulse is comprised of noise (incomplete mode-locking), the central peak (always proportional to Δv of the laser) is extremely narrow and is superimposed upon a broader peak corresponding to the pulse envelope. This broader peak has a contrast of only 2:1. Slight spatial misalignment can cause the central peak to disappear entirely (WEBER and DANIELMEYER, 1970). A multimode laser output in the absence of any mode-locking produces peaks with a contrast of 1.5:1. Thus, it is necessary to carefully determine the contrast ratio before a pulse-width can be deduced. It is also important to note that all information about pulse shape is lost in a measurement of $G^{(2)}(\tau)$. An estimate of pulse-width can only be made by assuming a pulse shape. The full width at half maximum (FWHM) $\Delta \tau$ of $G^{(2)}(\tau)$ is related to the pulse-width Δt (FWHM) by $\Delta \tau = \sqrt{2} \, \Delta t$ for a Gaussian pulse and $\Delta \tau = 2 \Delta t$ for a Lorentzian pulse. In order to determine actual pulse shape one has to resort to a measurement of higher-order correlation functions (BLAUNT and KLAUDER, 1969). Such measurements have been performed (ECKHARDT and LEE, 1969; RENTZEPIS et al., 1970; AUSTON, 1971) but are not always practical.

The SHG measurement procedure is illustrated in Fig. 3.2b. Here the pulse is divided into two beams that then travel collinearly in the nonlinear crystal with an adjustable time delay between them. If the two beams have the same polarization and τ is the time delay between them, this procedure is formally equivalent to a TPF experiment as described above. This type of SHG is called "first kind". A plot of second-harmonic output intensity versus τ yields precisely the same contrast ratios given for a trace of fluorescence versus distance in TPF. Whereas TPF has gained wide acceptance because of its adaptability to single pulse experiments, the equivalent SHG scheme has proven convenient for use with cw dye lasers. Several experimental results are given later.

A slightly different SHG measurement can be performed by giving the two beams opposite polarizations. It is then possible to choose a crystal orientation such that neither beam generates second-harmonic independently but signal is produced when overlap of the pulses exists. This is called SHG of the "second kind". The effect is one of removing the background from the plot of SHG versus τ and giving a direct trace of $G^{(2)}(\tau)$. One then expects a contrast ratio of $1:0$ for good pulses and a ratio of $2:1$ for pure noise. With such a scheme one might expect a less ambiguous measure of the pulse. In fact, without a background reference it can be more difficult to determine whether one has proper spatial alignment and is maintaining it over the scan. Because of crystal birefringence, the two orthogonally polarized beams also travel with different group velocities in the crystal. Thus, SHG experiments of the second kind may also require the use of very thin crystals.

The various mode-locking and pulse measurement techniques have all been applied to dye lasers. We now turn to discussion of results obtained with experimental dye-laser systems.

3.2. Pulsed Dye Lasers

In this section we discuss the generation of bursts of short pulses from mode-locked dye lasers. The laser systems described are pumped either with a flashlamp or another pulsed laser. Relatively high pumping intensities can be achieved in this way and the use of a large variety of dyes is possible. Overall time duration of the output is within the range 0.01–$10\,\mu$sec.

Early attempts at mode-locking an organic dye laser utilized the output of a mode-locked laser as an optical pump. As described in the previous section, pumping in this manner provides the modulation necessary for mode-locking by rapidly driving the gain medium in a periodic manner. Picosecond duration pump pulses can be provided at high powers by Nd:glass lasers ($\lambda = 1.06\,\mu$) and ruby lasers ($\lambda = 694.3$ nm). Frequency doubling these lasers gives similar pulses at 530 nm and 347.2 nm. Thus, the mode-locked laser pumping technique can be used to obtain ultra-short dye-laser pulses at a number of different wavelengths.

The first experiments used a mode-locked, frequency-doubled Nd:glass laser as the pump for rhodamine 6G and rhodamine B lasers (GLENN et al., 1968; SOFFER and LINN, 1968). The optical resonator of the dye laser was adjusted in length to provide sychronism with the pumping mode-locked pulses. When the length of the dye-laser resonator l_d is made equal to (or a submultiple of) the length l_p of the pumping resonator, trains of pulses are observed with periods equal to (or a submultiple of) the pump pulse period, providing the dye cell is placed at one end of the cavity. With the pumping and dye-laser resonators equal in length and the dye cell placed at intermediate harmonic positions $1/n$ for $n = 3$, 4 two optical pulses are observed for each pumping pulse. One of the pulses will be more intense than the other because the pulses traveling in opposite directions in the resonator reach the dye cell together only once per pump pulse. With the dye cell placed in the center of the dye-laser cavity and $l_d = (m/n) l_p$ where $m/n = 1$, 2/3, 1/2; $2(m/n)$ pulses are obtained for each pumping pulse. Other dye lasers which have been mode-locked in this manner include DTTC[1] (BASS and STEINFELD, 1968) and DTNDCT[1] (BRADLEY et al., 1968; BRADLEY and DURRANT, 1968) pumped by mode-locked ruby, dimethyl POPOP[1] (TOPP and RENTZEPIS, 1971) pumped with frequency-doubled ruby, and several infrared polymethine dyes (DERKACHYOVA et al., 1969) pumped at 1.06 μm with a Nd:glass laser. Wavelength tuning of dye lasers mode-locked this way has been achieved by the use of a rotatable grating as an end reflector of the resonator (SOFFER and LINN, 1968) and by mixing two laser dyes to shift the gain peak (GLENN et al., 1968).

In the initial observations of mode-locking the pulses were detected with photodiodes having a time response on the order of 0.5 nsec. This type of detector is sufficient to detect the pulse energy and repetition rate but cannot resolve the pulsewidth. The pulsewidths were then determined using the TPF technique described in the previous section. The pulsewidth for the Nd:glass-pumped rhodamine 6G dye laser was measured to be 10 psec (SOFFER and LINN, 1968). From this measurement the peak powers were estimated to be 10^6 watts for a pumping power of 10^8 watts.

A comment may be made about the rapid gain fluctuations one achieves with mode-locked pulse pumping. High gains are, of course, generated in a matter of picoseconds by a pump pulse of that duration. Relaxation of the gain may occur over a much longer time. The fluorescent lifetime of rhodamine 6G, for example, has been measured to be about 5×10^{-9} sec. In a laser, on the other hand, the intense

---

[1] DTNDCT = 1,1'-diethyl-γ-nitro-4,4'-dicarbocyanine tetrafluoroborate.
  DTTC   = 3,3'-diethylthiatricarbocyanine.
  POPOP = 1,4-di-[2-(5-phenoxazolyl)]-benzene.

internal powers drive the gain down by means of stimulated emission. At high enough powers, the gain duration and hence the dye-laser output pulsewidth can closely approximate the picosecond pumping duration. It has even been demonstrated experimentally that pico-second duration dye-laser pulses can be produced in traveling-wave fashion without the use of any resonator (Mack, 1969b).

Direct modulation of resonator loss has also been used to mode-lock a dye laser. With an acoustic modulator in the resonator of a flashlamp-pumped coumarin laser, mode-locked pulses with a width of several hundred picoseconds were generated at $\lambda = 460$ nm (Ferrar, 1969). The pulse train was a few microseconds in duration and the shortest pulses were observed at the end of the train. This gives some indication of the time it takes for pulses to develop with the slower modulation speed of such a system. Such a scheme is probably best used to generate pulses at otherwise unobtainable wavelengths.

The most interesting and successful approach to mode-locking ruby and Nd:glass laser systems has been passive mode-locking with a saturable absorber. Passive mode-locking of an organic dye laser was first reported by Schmidt and Schäfer (1968). They observed mode-locking of a flashlamp-pumped rhodamine 6G dye laser using an organic dye as a saturable absorber. The emission was seen to consist of a train of equally spaced pulses with a $c/2L$ repetition frequency of 1 GHz, but a pulsewidth determination in this early experiment was limited by the bandwidth of the photodiode and oscilloscope (0.4 nsec). The saturable absorber (DODCI)[2] used still remains one of the most effective mode-locking dyes for the rhodamine 6G dye laser. Bradley and O'Neill (1969) essentially reproduced these results with both rhodamine B and rhodamine 6G. They also observed sporadic self-mode-locking of the rhodamine 6G dye ostensibly due to the presence of the overlapping singlet absorption.

A diagram of a passively mode-locked dye laser is shown in Fig. 3.3. This is essentially the same configuration as that reported by Bradley (1971). The gain dye cell is pumped with a flashlamp. All surfaces are either at Brewster angle or wedged to prevent etalon resonances. The mode-locking dye solution is placed next to the mirror at the end of the resonator. This is found to be an optimum position (Bradley et al., 1969). The grating reflector is used for wavelength selectivity.

Measurements of pulsewidths in this system were made using the TPF technique. The authors (Bradley et al., 1969) reported that pulses as short as 6 psec could be inferred from two-photon-fluorescence. The observed contrast ratios were, however, considerably less than the

---

[2] DODCI = 3,3'-diethyloxadicarbocyanine iodide.

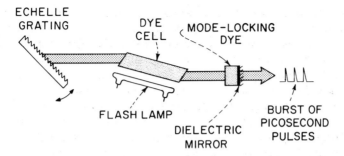

Fig. 3.3. Schematic of a flashlamp-pumped dye laser in configuration for mode-locking

expected 3:1. While imperfect alignment of the TPF apparatus and background fogging of the film might explain this, the possibility still existed that incomplete mode-locking might be the problem.

Although TPF is a valuable tool for estimating pulse widths, the use of linear, time-resolved measurements has been responsible for clearly demonstrating that well-isolated, ultra-short pulses are produced by mode-locked dye lasers. The ultrafast streak camera of BRADLEY et al. (1971b), capable of picosecond pulse resolution, was the first linear detector capable of time-resolving mode-locked dye-laser pulses. The ultrafast streak camera consists of a conventional streak imaging tube with a fine mesh high-potential extraction electrode placed near the photocathode to reduce the photoelectron transit time spread (BRADLEY et al., 1971a). The imaging tube was followed by an image intensifier and a camera. In Fig. 3.4 we show a streak recording of the output of a mode-locked dye laser reported by ARTHURS et al. (1972). The streak record in Fig. 3.4a shows a group of 4 pulses obtained when a delay line was arranged to produce, from a single laser pulse, pairs of sub-pulses separated by 6 psec, with single sub-pulses spaced 60 psec from each close pair. The microdensitometer trace of Fig. 3.4b shows the central pulse pair clearly resolved. The recorded pulse half-width was measured to be 2.5 psec. From the results of these and other measurements the system was found to have a 1.4 psec time dispersion limit for a minimum laser pulse of 1 psec.

The ultrafast streak camera has been applied to the measurement of background intensity of picosecond pulses from a passively mode-locked rhodamine 6G dye laser (BRADLEY et al., 1971b). Even though tube saturation prevented high resolution within several hundred picoseconds of the pulse, the ratio of peak pulse power to that of the background intensity was determined to be greater than 10^4. This implied that less than 3 percent of the laser energy output occurred between

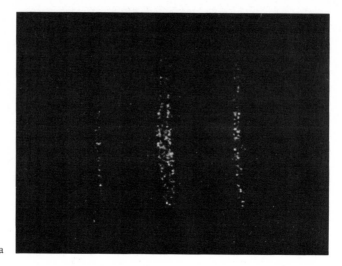

a

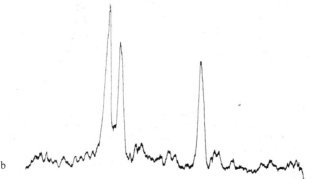

b

Fig. 3.4. a Streak camera record of a mode-locked dye-laser pulse. A delay line was arranged to produce, from a single pulse, pairs of sub-pulses separated by 6.6 psec with a single sub-pulse spaced 60 psec from each pair. b Microdensitometer trace of the above photograph (From Arthurs et al., 1972).

pulses and that the pulse power was 20 MW. The streak camera has also been used to study the time development of a mode-locked pulse from initial noise in a dye-laser cavity (Arthurs et al., 1972a). In a flashlamp pumped dye-laser resonator a picosecond pulse can be developed in about 15 round trips.

High power mode-locked lasers typically produce pulses whose spectral content is not directly related to pulsewidth. Often for pico-second pulses with durations less than 5 psec, the spectral bandwidth (3–10 nm) is up to 100 times greater than the Fourier-transform limit

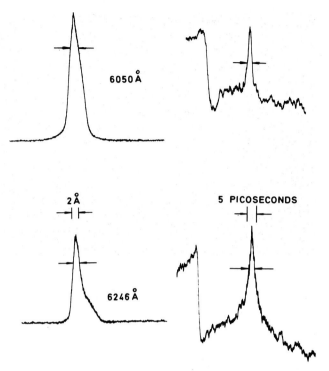

Fig. 3.5. Microdensitometer traces of simultaneously recorded pulse spectra and TPF tracks indicating approximately transform-limited pulses. Recordings made at 605 nm and 624.6 nm, the opposite ends of the picosecond pulse tuning range. Time calibration marks for the TPF tracks given for an assumed Lorentzian shape. (Multiply by $\sqrt{2}$ for Gaussian shape.) Left hand sides of TPF traces indicate $\times 2.4$ intensity steps (From ARTHURS et al., 1971)

set by the pulse durations (BRADLEY et al., 1969). In Nd:glass lasers there have been reports of sub-picosecond structure, partial linear frequency sweeps, and self-phase modulation which indicates that a degree of both amplitude and frequency modulation occur in these pulses. ARTHURS et al. (1971) reported that by using an interferometer filter internal to the resonator it was possible to reduce the bandwidth of a mode-locked dye laser so that the output corresponded to the Fourier-transform limit set by the pulse width. Tunable transform limited pulses were obtained over a wavelength range of 23 nm from 602–625 nm. Over the tuning range the pulse bandwidth was about 0.2 nm and the pulses were from 2–4 psec. In Fig. 3.5 are shown the microdensitometer traces of TPF patterns and frequency spectra for the two extreme wavelengths of operation. The time calibration tracks are for an assumed Lorentzian pulse shape. Best results were obtained near threshold. At powers far above threshold the spectrum broadened. The broadening was observed to increase with time in the pulse train suggesting that the broadening was due to self-phase modulation intensified by multiple passes through the laser medium. Good single pulses may be obtained by electro-optically switching a pulse out of the train before broadening is significant.

The wavelength range of passively mode-locked lasers is in principle limited only by the availability of laser dyes and absorbers. ARTHURS et al. (1972b) have demonstrated tunable picosecond pulse generation over the range 584 nm to 704 nm. Various combinations of three laser dyes and four polymethine saturable absorbers were used to cover this range. The broadest spectrum covered by a single pair was 652 nm to 704 nm for a gain medium mixture of rhodamine 6G and cresyl violet and DTDCI[3] absorber. Tuning was accomplished with a narrow-gap (4–8 μ) interferometer inside the resonator. To a certain extent the range in each case was limited by the free spectral range of the interferometer. Pulse widths were less than 6 psec over the entire range.

For purposes of summary the wavelength range covered at present by pulsed, mode-locked dye lasers is shown in Fig. 3.6. Literature references for each system are included on the figure. Although the peak mode-locked powers from dye lasers remains less than that ob-

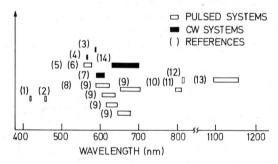

Fig. 3.6. Wavelength ranges over which dye laser mode-locking has been reported for both pulsed and cw systems. The numbers in brackets indicate corresponding references

(1) Topp and Rentzepis (1971).	(8) Schmidt and Schäfer (1968).
(2) Ferrar (1969b).	(9) Arthurs et al. (1972b).
(3) Kuizenga (1971).	(10) Bradley et al. (1968).
(4) Dienes et al. (1971a).	(11) Bradley and Durrant (1968).
(5) Glenn et al. (1968).	(12) Bass and Steinfeld (1968).
(6) Soffer and Linn (1968).	(13) Dechachyova et al. (1969).
(7) Ippen et al. (1972).	(14) Runge (1972).

[3] DTDCI = 3,3′-diethylthiadicarbocyanine iodide.

tainable from, for example, Nd:glass lasers, pulses of a few picoseconds duration, and powers of 20 MW can be obtained. By means of a series of dye-laser amplifiers, pulses can be amplified into the gigawatt range (BRADLEY, 1971; BRADLEY et al., 1971 b).

3.3. Continuous Dye Lasers

With the development of the cw dye laser it was apparent that continuous trains of mode-locked picosecond pulses were a possibility. Although single picosecond pulses had proven themselves to be useful tools for picosecond spectroscopy, continuous trains of pulses seemed to point the way to a new dimension in measurement possibilities.

The first attempt at mode-locking the cw dye laser utilized direct modulation of the resonator loss. DIENES et al. (1971 a) reported on a stable train of mode-locked pulses that was obtained with an acousto-optic loss modulator in the dye-laser resonator. Mode-locking was achieved with a modulation depth of several percent at 75 MHz, the fundamental cavity spacing, as well as at 225 MHz, the third harmonic of the mode spacing.

In Fig. 3.7 we show a measurement system for analyzing cw mode-locked pulses from a dye laser. A portion of the pulse is directed to a diode detector and a sampling scope to observe the pulses with a 100 psec resolution. A scanning Fabry-Perot allows an instantaneous display of the oscillating spectrum. The pulse train is sent into an interferometer where it is split into two parts with a variable delay between the

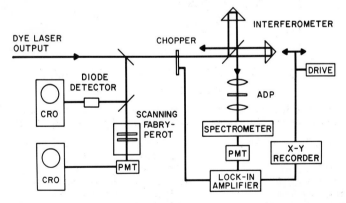

Fig. 3.7. Schematic of apparatus used to monitor and measure pulses from a cw mode-locked dye laser. The photodiode and sampling scope determine repetition rate and pulse energy. A scanning Fabry-Perot monitors the oscillating spectrum. SHG auto-correlation in ADP provides measure of pulse width

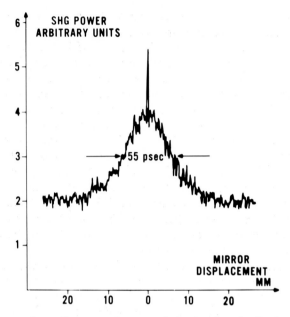

Fig. 3.8. The experimentally measured autocorrelation function of pulses from an actively mode-locked cw dye laser. Contrast ratios indicate that mode-locking is incomplete. The pulse envelope of 55 psec is determined assuming a Gaussian shape

two determined by the length of one arm of the interferometer. The output of the interferometer is directed into a ADP crystal to generate the second harmonic. In the measurements described here second-harmonic generation of the first kind, in which the polarizations of both beams from the interferometer are parallel, was chosen to permit phase matching at the dye-laser wavelength with available crystals.

Figure 3.8 exhibits a trace of the second harmonic intensity vs. interferometer mirror position. The contrast ratios observed here agree very closely with the theoretical predictions described in the first section. The curve consists of a sharp central spike on top of a broader peak and a flat background. Such a curve is characteristic of partial mode-locking and the expected contrast ratios for these three features are 3:2:1. The width of the central spike indicates the coherence of the laser and is inversely proportional to the full spectral bandwidth. The broader peak is a measure of the width of the optical pulse envelope. Assuming a Gaussian pulse shape, the half-power width of the mode-locked pulse is 55 psec. The curve in Fig. 3.8 was taken with 225 MHz modulation. At 75 MHz the pulse envelope was observed to broaden to about 70 psec. At the scan rate shown the central peak was not entirely resolved, but it could be resolved at slower rates.

For reasons not yet understood the pulsewidths measured for active mode-locking were an order of magnitude longer than one would expect from the theory of KUIZENGA and SIEGMAN (1970) for mode-locked homogeneously broadened lasers. In addition there appeared to be no relationship between the spectral broadening which was observed to occur during mode-locking and the measured pulsewidths.

Active mode-locking has also been attempted with a phase modulator (KUIZENGA, 1971). As with loss modulation, mode-locking in this system remained incomplete and relatively long pulses were obtained. Pulses about 500 psec in width with considerable substructure were observed at a modulation frequency of 312 MHz. With the help of a thin internal etalon, single pulses of 200 psec were obtained. Stable mode-locking, however, appears to be more difficult with phase modulation than with amplitude modulation.

Mode-locking with a continuous mode-locked pump source has also been achieved. RUNGE (1971) placed a laser dye internal to a He–Ne laser resonator operating at 632.8 nm. The dye acted both as laser medium and as saturable absorber to mode-lock the He–Ne laser. The dye lased synchronously with the He–Ne laser in the same resonator. In the first experiments of this type, laser action was limited to near strongly dispersive Ne transitions. The dispersion of the Ne transitions acted to synchronize the 632.8 nm pumping pulse and the dye-laser pulse. Independent tuning of the mode-locked dye-laser emission was achieved by introducing a prism into the resonator and providing a separate resonator length adjustment for both frequencies (RUNGE, 1972). The mode-locked dye-laser pulses tended to replicate the He–Ne pulses which were 270 psec. Mode-locking was observed with cresyl violet, nile blue A, resazurin, DODCI, and DTDC.

An actively mode-locked argon laser has been used to pump a rhodamine 6G cw dye laser with resonator lengths adjusted for synchronism (DIENES et al., 1971 b). Pulses from the dye laser were observed to be somewhat shorter than the pumping pulses but were about 125 psec in length. While active mode-locking may be useful for some applications, ultrashort pulses in the picosecond range have not been obtained.

Recently, passive mode-locking of the cw rhodamine 6G dye laser has been reported (SHANK et al., 1972; IPPEN et al., 1972b; O'NEILL, 1972). Continuous trains of pulses as short as 1.5 psec were observed. The experimental optical resonator is shown in Fig. 3.9. The laser utilizes a five-mirror arrangement that is an extension of a previously described resonator with astigmatic compensation (KOGELNIK et al., 1972b;

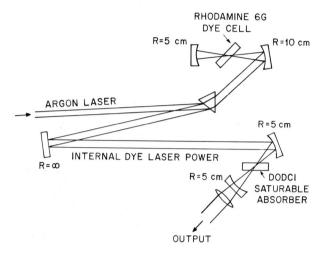

Fig. 3.9. Schematic of the resonator configuration used to passively mode-lock a cw dye laser

DIENES et al., 1972). The 514.5 nm output of a continuous argon laser is coupled into the resonator through a quartz prism and is focused by a mirror into a cell containing rhodamine 6 G. At the other end of the dye-laser resonator is a similar cell containing a $2 \times 10^{-5}\,M$ solution of DODCI in methanol. Both cells are 1 mm thick and oriented at Brewster's angle. The dye laser mode has a diameter of about 25 μm in the rhodamine 6 G cell and a smaller diameter of about 15 μm in the saturable absorber. The overall resonator length of 2 meters corresponds to a round trip time of 12 nsec.

Without any saturable absorber the dye laser operated continuously with a bandwidth of 0.1 nm. The laser could be tuned by rotating the flat mirror. With the addition of a saturable absorber, mode-locking was observed by the appearance of pulses on the sampling oscilloscope. Mode-locking was always accompanied by considerable broadening of the spectrum. The correspondence between the oscillating spectrum and inverse pulsewidth was such that, after calibration the spectrum alone was a good monitor of the pulsewidth.

The actual pulsewidths were determined from autocorrelation measurements using second-harmonic generation. In Fig. 3.10 we show two experimental traces of second-harmonic intensity vs. interferometer position. The difference between the two traces is a slight cavity adjustment. A considerably slower scan rate was used in Fig. 3.10b to ensure resolution. The 2.3 and 1.5 psec widths are calculated assuming Gaussian pulse shapes. If the pulses are Lorentzian shaped, their widths

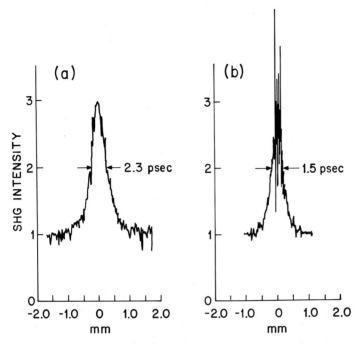

Fig. 3.10. Experimentally measured pulse autocorrelation functions for the passively mode-locked cw dye laser. The difference between the two traces is a slight resonator adjustment. The scan rate for (b) was slower than for (a) and begins to reveal wavelength fringes of the interferometer. Pulsewidths calculated for a Gaussian profile. (Divide by $\sqrt{2}$ for a Lorentzian)

are 1.7 and 1.1 psec, respectively. The observed contrast ratios of the pulse to background compare well with the theoretically expected value of 3:1. The extreme amplitude fluctuations near the center of the trace in Fig. 3.10b were not due to noise but to unresolved wavelength fringes of the interferometer. Stability of the pulse train is evidenced by the fact that one of these measurements made over a period of 2 min yields an average over 10^{10} pulses. The frequency spectra corresponding to the correlation traces in Fig. 3.10a and Fig. 3.10b indicated oscillation bandwidths (FWHM) of 0.5 and 0.65 nm, respectively. These results can imply either spectrally limited Lorentzian-shaped pulses or Gaussian pulses within a factor of 2 of the spectral limit.

The peak power in the pulses inferred from average power and pulsewidth measurements is in the range 50–100 watts. Best mode-locking occurs in this laser not too far above threshold. At higher powers it was difficult to suppress multiple-pulse operation (more than

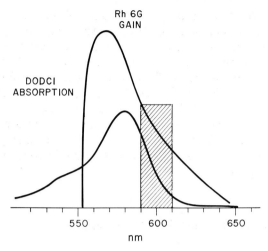

Fig. 3.11. The gain curve for rhodamine 6G and the low intensity absorption spectrum of DODCI. (Linear vertical scale in arbitrary units.) The crosshatched area indicates the wavelength range for mode-locking under continuous operation

one pulse in the resonator). This is due to the recovery of the laser gain in a period short compared to a round-trip time of the resonator.

In Fig. 3.11 we display a plot of rhodamine 6G gain and the low intensity DODCI absorption as a function of wavelength. The range in which a short pulse mode-locking has been observed is shown in the shaded box, and extends from 590–610 nm. Most of this tunability was obtained with a single saturable-absorber concentration, and varying the concentration did not affect the range significantly. Note that mode-locking occurs only on the long wavelength side of both the gain and absorption maxima. It has been suggested that this behavior might result from the formation of new isomeric species in the absorber dyes and a corresponding shift in absorption to the red. In support of this hypothesis we have observed a novel bistable mode of cw operation with the absorber in the resonator. As the laser is tuned from the mode-locking regime toward the yellow, strong cw oscillation becomes dominant. Furthermore, this oscillation is bistable. If the oscillation is interrupted it will not reinitiate. The laser must be tuned from red to the yellow to obtain oscillation in the bistable region. This behavior could indicate that the laser radiation is required to convert the absorber molecules to an isomeric configuration with reduced yellow absorption. In fact, such a reduction in the yellow absorption of DODCI has been observed in measurements of the absorption constant under intense illumination (IPPEN and SHANK, 1972c). The peak absorption under such illumination is observed to shift to about 600 nm.

Finally the question arises as to the mechanism at work in the saturable absorber that explains the short pulse behavior. Measurements of fluorescence lifetimes in DODCI have yielded values from 0.3 nsec (BRADLEY et al., 1969) to 1.3 nsec (IPPEN and SHANK, 1972c) and 1.6 nsec (CIRKEL et al., 1972). It is not fully understood how such a long relaxation time can produce 1.5 psec pulses. There is a possibility that a fast lifetime could be present in the system which as yet remains unidentified. More discussion on possible mode-locking mechanisms will be presented in the next section.

3.4. Discussion of Short Pulse Formation

In this section we are going to discuss some of the aspects of short pulse formation as applied to mode-locking of organic dye lasers. The literature on picosecond pulse formation has primarily been concerned with the development of short pulses in Nd:glass and ruby lasers. The understanding of passive mode-locking in organic dye lasers still remains to be put on a firm basis.

For extensive discussions of pulse formation in Nd:glass and ruby lasers the reader is referred to papers by KRYUKOV and LETOKHOV (1970) and FLECK (1968). In general, for these lasers the mode-locking process is viewed as one in which the passive nonlinear absorber strongly attenuates low intensity radiation and selectively passes short bursts of high intensity. Such bursts of high intensity are always present as initial noise fluctuations in the laser. Because the starting conditions are subject to these fluctuations, the resulting output pulses may also exhibit statistical behavior. In particular, the appearance of incompletely suppressed sub-pulses is not uncommon. In the rate equation approximation, the output pulsewidths are limited by the absorber lifetime but can depend also on the initial noise burst. FLECK (1968) has shown that statistically starting fluctuations are removed with time and that a laser operating near threshold eventually produces pulses whose width is the saturable absorber lifetime. For the case of these solid-state lasers the upper-state lifetime of the gain medium is so slow that nonlinear amplification has very little effect on the pulse shape. It is generally observed that the shortest pulses appear near the beginning of a train where the saturable absorber is most effective. With increasing number of passes in the resonator, high power pulses may be both spectrally broadened and broadened in time.

In organic dye lasers the situation is somewhat different. First, the upper state lifetime of the gain medium is on the order of the round-trip time and gain recovery is possible between pulses. Thus, gain saturation can be expected to have an influence on the mode-locking process. Second, as stated in the last section, the fluorescent lifetime of the saturable absorber appears to be much longer than the pulse widths observed. Of course, it is possible that there is a fast recovery time in the family of nonlinear absorbers led for mode-locking dye lasers but this has yet to be observed experimentally.

Let us now consider how a nonlinear saturable absorber interacts with a pulse which is much shorter than the upper state lifetime. Unlike the case of long pulse saturation ($\tau_p \gg T_1$) where pulse intensity determines the saturation level, a short pulse ($\tau_p \ll T_1$) saturates with pulse energy. That is, with no recovery during the pulse the absorber can only absorb a given number of photons. Then it remains bleached for a time T_1. The saturation energy in photons is given by (Kryukov and Letokhov, 1970)

$$E_s = 1/2\sigma \tag{3.3}$$

where σ is the absorber cross-section. Saturation of this type can only affect the leading edge of the pulse. If there is no other pulse forming mechanism, gain over the entire pulse will gradually cause the trailing edge of the pulse to lengthen until it is limited by absorber recovery.

A nonlinear amplifier can shape the pulse by preferentially amplifying the leading edge of the pulse. This can have the effect of narrowing the pulse. A necessary condition for narrowing is that the pulse have a relatively steep leading edge. A slow leading edge can result in a forward displacement of the pulse peak without any reduction in width. Gaussian-shaped pulses and pulses with a steplike front edge are examples of pulses which satisfy the narrowing criteria. It is assumed that the nonlinearity is saturation according to (3.3) with σ the amplifier emission cross-section.

A two component laser consisting of both a nonlinear amplifier and a saturable absorber can narrow a pulse of arbitrary shape under the right conditions. If a pulse propagates in a two component medium with the absorber cross-section much larger than the amplifying cross-section, a contraction of the pulse duration takes place independent of the input wave form (Kryukov and Letokhov, 1970). The nonlinear absorber acts to continuously increase the slope of the leading edge while amplifier saturation discriminates against the trailing edge. The requirement on the cross-sections is equivalent to saying that the nonlinear absorber must saturate with less energy than the nonlinear amplifier. This is necessary for the peak of the pulse to achieve greater net gain than either the leading or trailing edges. It follows that a steady state situation can be achieved in which the total pulse energy remains the same from pass to pass and the pulse is indefinitely shortened. In a continuous laser, pulse broadening mechanisms not included in the above model may determine the ultimate pulse width. It is important, however, that ultra-short pulses can be formed by the combination of a saturable absorber and saturable gain medium each of which may have relatively long recovery times.

In an optical resonator the pulse round-trip can also be an important factor. Most passively mode-locked lasers have had a round-trip time that greatly exceeded the apparent recovery time of the absorber. If absorber recovery between pulses is not complete, there results a continuous saturation which reduces the effect of the absorber. A DODCI mode-locked Rh 6G laser has, however, been successfully operated with a round-trip time as short as 1 nsec (SCHMIDT and SCHÄFER, 1968). A factor that may be most important to mode-locking in dye lasers is the ratio of round-trip time to gain recovery time. The stability of the indefinite pulse narrowing process described above depends upon this ratio as well as the other gain and loss parameters. Stability requires a net loss on both the leading and trailing edges of the steady state pulse. The first analysis of mode-locking based upon these steady state stability criteria has been presented recently by NEW (1972). A qualitative explanation of the existence of a stability regime is the following: At the beginning of oscillation, gain within the resonator must exceed loss. By the time steady is reached, loss for the leading edge of the pulse must exceed the gain. Thus it is necessary that the amplifier must not completely recover in a round-trip time. This sets an upper limit on the length of the resonator. On the other hand, if the resonator is too short there is insufficient time for gain recovery and nonlinear amplification ceases to play a role in pulse shaping. The leeway that one has with resonator length will be determined by the initial gain and loss values and the cross-section ratio. If all the appropriate conditions are not met, mode-locking may result only in relatively long pulses or in multiple pulse behavior.

The foregoing discussion may yield an intuitive basis for understanding pulse formation but has not answered the questions of the ultimate pulsewidths and energies obtainable from a mode-locked organic dye laser.

4. Structure and Properties of Laser Dyes

K. H. DREXHAGE

With 6 Figures

An essential constituent of any laser is the amplifying medium which, in the context of this book, is a solution of an organic dye. Since the beginning, the development of the dye laser has been closely tied in with the discovery of new and better laser dyes. The phthalocyanine solution employed for the original dye laser (SOROKIN and LANKARD, 1966) is hardly used today, but the compound rhodamine 6G, found soon afterwards (SOROKIN et al., 1967), is probably the most widely employed laser dye at the present time. In the years following the discovery of the dye laser various other compounds were reported for this purpose. Almost all were found by screening commercially available chemicals, but this source of new laser dyes soon became exhausted. It was reported in 1969 that a survey of approximately one thousand commercial dyes showed only four to be useful (GREGG and THOMAS, 1969), and three of these belonged to classes of laser dyes that were already well known. Considering the large number of available chemicals, it is perhaps surprising that so few good laser dyes have been found so far. The reason for this is that some very special requirements must be met by such dyes and this excludes the majority of organic compounds. This chapter is intended to give the reader some insight into the relations between molecular structure and the lasing properties of organic dyes, relations which have recently been applied in the planned synthesis of new laser dyes. In addition, the physicochemical properties of the most important laser dyes are discussed here in some detail.

4.1. Physical Properties of Laser Dyes

Organic dyes are characterized by a strong absorption band in the visible region of the electromagnetic spectrum. Such a property is found only in organic compounds which contain an extended system of conjugated bonds, i.e., alternating single and double bonds. Whereas the light absorption of dyes cannot be derived *rigorously* from their molecular structure owing to the complexity of the quantum-mechanical problem, more or less simple models have been found that are capable of explaining many experimental observations, at least within a given class of dyes. By adjusting the model parameters on the basis of empirical data, it is possible to predict the absorption properties of yet unknown dyes. For a more detailed discussion, the reader is referred to Chapter 1. The nonradiative processes in dye molecules (which are of the utmost importance in dye lasers, as discussed below) are even less amenable than light absorption to an accurate quantum-mechanical treatment.

The long-wavelength absorption band of dyes is attributed to the transition from the electronic ground state S_0 to the first excited singlet state S_1. The transition moment for this process is typically very large, thus giving rise to an absorption band with an oscillator strength on the order of unity. The reverse process $S_1 \rightarrow S_0$ is responsible for the spontaneous emission known as fluorescence and for the stimulated emission in dye lasers. Because of the large transition moment the rate of spontaneous emission is rather high (radiative lifetime on the order of nanoseconds) and the gain of a dye laser may exceed that of solid-state lasers by several orders of magnitude.

When the dye laser is pumped with an intense light source (flashlamp or laser), the dye molecules are excited typically to some higher level in the singlet manifold, from which they relax within picoseconds to the lowest vibronic level of S_1, i.e., the upper lasing level (Fig. 4.1). For optimal lasing efficiency it would be desirable for the dye molecules to remain in this level until they are called on for stimulated emission.

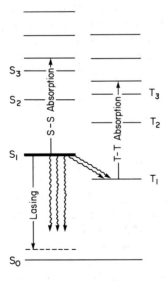

Fig. 4.1. Energy level diagram of an organic dye molecule

However, there are many nonradiative processes that can compete effectively with the light emission and thus reduce the fluorescence efficiency to a degree that depends in a complicated fashion on the molecular structure of the dye. These nonradiative processes can be grouped into those that cause a direct relaxation to the ground state S_0 (internal conversion) and those that are responsible for intersystem crossing to the triplet manifold. Because of the relatively long lifetime of the triplet molecules (in the order of microseconds) the dye accumulates during the pumping process in the triplet state T_1, which often has considerable absorption for the laser light. Thus not only are some of the dye molecules taken away from the lasing process but, owing to the triplet-triplet absorption, they cause an additional loss in the laser. The nonradiative decay to the ground state is comparatively less harmful for laser action. Nevertheless, in the ideal laser dye both processes should be negligible, so that the quantum yield of fluorescence has the highest possible value, 100%. In addition to these general requirements, an efficient laser dye in its first excited singlet state should have negligible absorption at the wavelength of the pump light and of the laser emission as well. Otherwise losses would occur, as in triplet-triplet absorption, because the decay to the first excited singlet or triplet level is nonradiative.

In addition to the aforementioned optical properties, the laser dye should have an absorption spectrum which matches the spectral distribution of the pump source. Since a substantial part of the light energy emitted by flashlamps is in the ultraviolet region, only dyes with moderate to strong absorption throughout this region can take full advantage of the pump light. If, on the other hand, the pump source is a laser with monochromatic emission, the dye should have a strong absorption at this wavelength. Although, in principle, a small absorption coefficient can be compensated for with a higher concentration, this is often not desirable because it also increases the absorption at the lasing wavelength, thus enhancing the cavity losses. In order to achieve a broad tuning range, dyes are required that have an unusually wide fluorescence band, or mixtures of dyes that absorb at the same wavelength but fluoresce with different Stokes shifts. Fluorescent dyes that react in the excited state to form a fluorescent product that is unstable in the ground state can be considered ideal for this purpose (see Subsection 4.5.5). Here the concentrations may be adjusted so that the gain in the regions of fluorescence of the original dye and its reaction product are approximately equal. Following the emission of light, the reaction product dissociates immediately and no additional absorption interfering with the lasing process is encountered, such as is likely to be the case in a mixed solution of several dyes.

Inevitably during the lasing process a certain amount of thermal energy is released, giving rise to temperature gradients in the solution that may cause optical inhomogeneities. In this respect, the best media appear to be water and its deuterated relative, heavy water. Therefore the ideal laser dye should be soluble in water and still maintain its lasing properties, which generally means that it must not form dimers at the lasing concentrations or that it responds readily to disaggregating agents (see Subsection 4.5.4).

Finally, we should like to mention photochemical stability as a property relevant to laser dyes. Whereas this is of lesser importance in dye lasers employing liquid solutions, the poor stability of many laser dyes is a serious problem in lasers where the dye is incorporated in a solid matrix and therefore cannot be circulated.

4.2. Internal Conversion $S_1 \rightarrow S_0$

The nonradiative decay of the lowest excited singlet state S_1 directly to the ground state S_0 is mostly responsible for the loss of fluorescence efficiency in organic dyes. Depending on the molecular structure of the dye and the properties of the solvent, the rate of the relaxation can vary by many orders of magnitude. Since there are several different structural features that contribute to the nonradiative decay $S_1 \rightarrow S_0$, the relation between molecular structure and fluorescence efficiency can be puzzling. For a general survey the reader is referred to [FÖRSTER, 1951 (p. 94); WEHRY, 1967; PARKER, 1968 (p. 428); BIRKS, 1970 (p. 142)]. In this section we dwell on some processes that are especially relevant to laser dyes.

4.2.1. Structural Mobility

It has been known for a long time that a rigid, planar molecular structure favors high fluorescence efficiency, as may be illustrated by the dyes phenolphthalein and fluorescein [FÖRSTER, 1951 (p. 109)]. Whereas the former is practically nonfluorescent[1] in alkaline solution, the introduction of the oxygen bridge causes the fluorescence quantum yield to assume the value of 90% in fluorescein. A similarly strong effect has

---

[1] Since there is always a finite probability of light emission by a molecule in an excited state, a compound can never be truly nonfluorescent. Even with very weakly fluorescing compounds the fluorescence can be observed when extremely intense light sources are used for excitation. Likewise, a quantum efficiency of 100% cannot be realized because the rate of nonradiative processes cannot be zero.

been observed in a number of other cases. In the laser dye rhodamine B the fluorescence efficiency in ethanol is dependent on the temperature (Huth et al., 1969). About 40% at 25 °C, the quantum yield increases to nearly 100% if the temperature is lowered, but it drops to only a few percent in boiling ethanol. These effects can be attributed to the mobility of the diethylamino groups, which is enhanced with increasing temperature. If these groups are rigidized, as in rhodamine 101, the quantum yield of fluorescence is found to be virtually 100%, independent of the temperature (Drexhage, 1972, 1973a; Drexhage and Reynolds, 1972). Still another structural mobility that increases the rate of internal conversion is illustrated by the dyes rosamine 4 and rhodamine 110. In the latter, as in other rhodamine dyes, the carboxyphenyl substituent is held in a position nearly perpendicular to the xanthene chromophore by the bulky carboxyl group, thus allowing little, if any, mobility. In the case of rosamine 4, however, the phenyl substituent can pivot to a certain degree, thus reducing the fluorescence efficiency in ethanol at 25 °C which is 85% in rhodamine 110, to 60% (Drexhage, 1972, 1973b). The fluorescence efficiency of rosamine 4 is also temperature-dependent, but to a lesser degree than in rhodamine B.

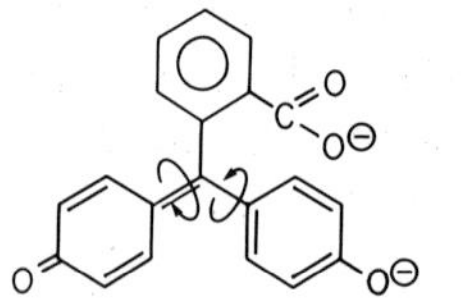

phenolphthalein

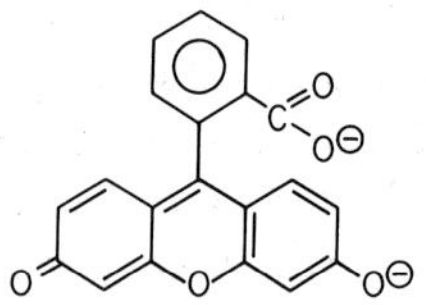

fluorescein

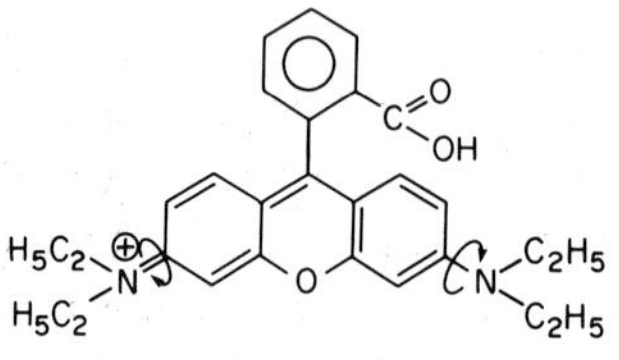

rhodamine B

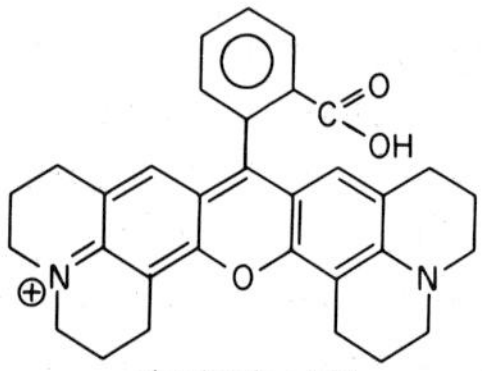

rhodamine 101

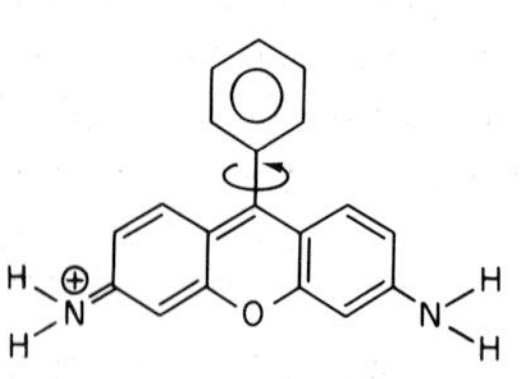

rosamine 4

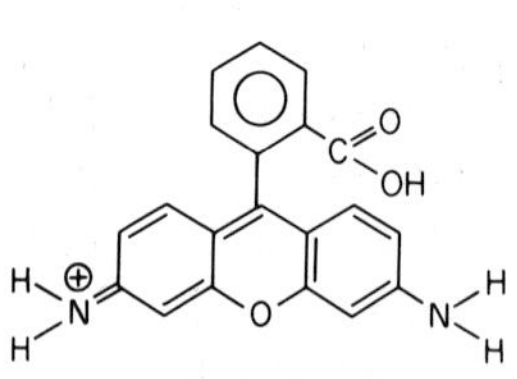

rhodamine 110

One must not conclude from the foregoing discussion that any dye with a completely rigid structure will have a very high fluorescence efficiency. The dye may, in fact, be nonfluorescent if one of the quenching processes to be discussed below is operative. On the other hand, it is also possible for a dye with a nonrigid structure to be highly fluorescent. This is the case, for instance, with rhodamine 110 and rhodamine 6G which in alcoholic solution have a fluorescence quantum yield of 85% and 95%, respectively, independent of temperature, although their amino groups are potentially mobile as in the case of rhodamine B. The reason for this different behavior lies in the excited-state π-electron density within the $C = N$ bond, which should be dependent on the number of electron-donating alkyl substituents (DREXHAGE, 1972, 1973a). If the π-electron density is high, the thermal energy of the solvent molecules is not sufficient to twist the amino groups out of planarity (rhodamine 110, rhodamine 6G). It is interesting that rhodamine B has a quantum efficiency of nearly 100% in viscous solvents like glycerol. This suggests that the chromophore is fully rigid in the ground state and loosens up only after excitation, provided the solvent is of low viscosity. In glycerol the viscosity is sufficiently high to prevent thermal equilibrium being reached during the radiative lifetime of a few nanoseconds. Hence the planarity of the ground state is not lost before light emission takes place. The fluorescence efficiency of rhodamine B and related dyes with dialkylamino end groups depends also in a peculiar way on solvent properties other than viscosity, and is moreover sensitive to other influences, as will be discussed in Section 4.7.

4.2.2. Hydrogen Vibrations

A pathway of internal conversion, which may be distinguished from the processes discussed above, can occur in certain dyes, even if their chromophore is fully rigid and planar. In contrast to other relaxation processes, this one is, in first approximation, independent of temperature and solvent viscosity. It involves the conversion of the lowest vibronic level of the excited state S_1 to a higher vibronic level of the ground state S_0, which then rapidly relaxes to the lowest vibronic level of S_0. The probability for this process is inversely proportional to the change in vibronic quantum number during the conversion. Because of the comparatively small mass of the hydrogen atom, the quanta of hydrogen stretching vibrations have the highest energies in organic dyes, and thus hydrogen vibrations are very likely to contribute to the mechanism considered here.

It can be expected that only those hydrogen atoms that are directly attached to the chromophore of the dye will influence the nonradiative process $S_1 \rightarrow S_0$. Furthermore, this mechanism should become increasingly effective with decreasing energy difference between S_1 and S_0. On the other hand, a replacement of hydrogen by deuterium should reduce the rate of nonradiative decay by this mechanism, and thus increase the fluorescence efficiency. This was indeed observed, for instance, in the dyes rhodamine 110 and Cresyl Violet (DREXHAGE, 1972, 1973 a). When these dyes were dissolved in monodeuterated methanol (CH_3OD), their quantum yields of fluorescence increased from 85 to 92% and from 70 to 90%, respectively, compared with the yields when normal methanol was the solvent. Since in CH_3OD solution the protons of the amino groups are readily exchanged with deuterons from the solvent, which is in large excess, the dye molecules exist here with deuterated amino groups. Among other xanthene dyes, the fluorescence quantum yield of fluorescein in alkaline alcoholic solution is 10% below and that of rhodamine 6G in alcohol 5% below the maximum value of 100%. As with rhodamine 110, these effects can be ascribed for the most part to hydrogen vibrations in the end groups of the chromophores, involving solvent molecules in the case of fluorescein. This mechanism, though of little importance in dyes that fluoresce in the visible range, can be expected to seriously reduce the fluorescence efficiency of infrared dyes.

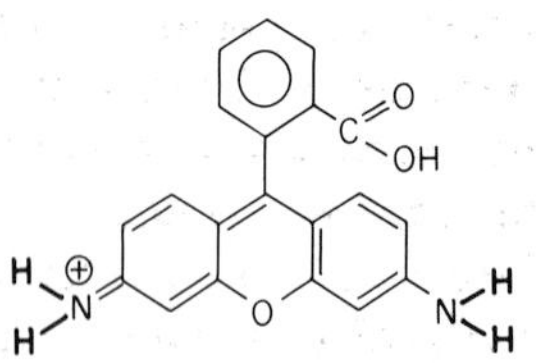

rhodamine 110

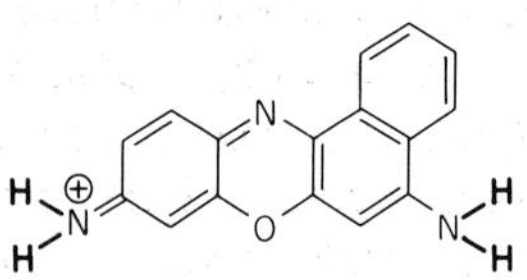

cresyl violet

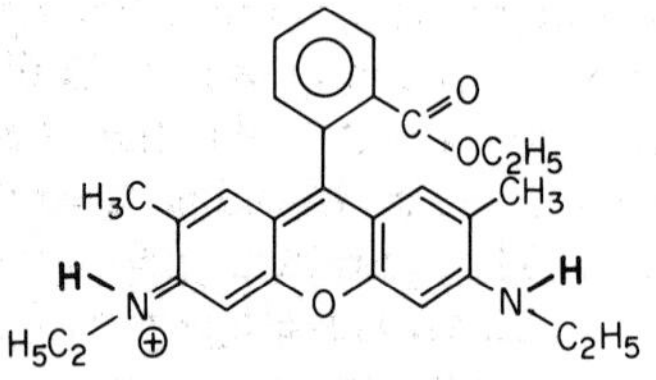

rhodamine 6G

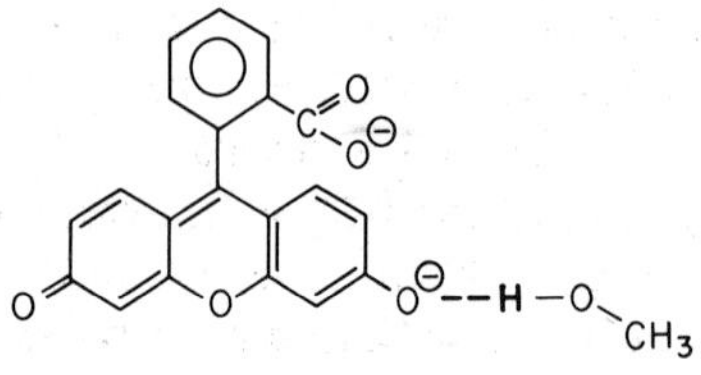

fluorescein

4.2.3. Other Intramolecular Quenching Processes

In addition to the mechanisms of internal conversion discussed already, there are several other intramolecular processes that may cause quenching of fluorescence. For instance, if a part of the dye molecule is strongly electron-donating or withdrawing, a reversible charge transfer may occur between this group and the excited chromophore, resulting in the loss of electronic excitation. Thus when a nitro group is introduced into the carboxyphenyl substituent of a rhodamine dye, a nonfluorescent derivative is obtained. Likewise, a substituent with a low-lying singlet or triplet state may quench the fluorescence via energy transfer. Furthermore, it is possible that under certain circumstances the intersystem crossing process $S_1 \rightarrow T_1$ effectively drains the state S_1 before the emission of fluorescence. Apart from the processes discussed in Section 4.3, this is observed with molecules where a $n - \pi^*$ transition occurs at lower energy than the $\pi - \pi^*$ transition of the main absorption band. Since these processes are at present of little importance with regard to laser dyes, they are not treated here in detail.

4.2.4. Determination of Fluorescence Yields

The fact that a great variety of nonradiative relaxation processes may compete with the emission of fluorescence can make it very difficult to elucidate the relations between molecular structure and the fluorescence quantum yield of organic dyes. The problem is aggravated by the many pitfalls encountered in the measurement of quantum yields. It has proven to be immensely difficult to determine an absolute quantum yield with an accuracy of only $\pm$ 5%. Most authors have overestimated the accuracy of their methods, even after careful evaluation of potential errors. A recent review of the measurement of fluorescence quantum yields gives an excellent account of the problems (DEMAS and CROSBY, 1971).

Easier than an absolute determination is the measurement of fluorescence efficiency relative to that of a standard solution. For dyes that fluoresce in the center of the visible region, one may use an alkaline fluorescein solution whose fluorescence quantum yield has been determined carefully with several independent methods and is generally assumed now to have the value 90 $\pm$ 5%. The quantum yield values given by me are based on this figure. In an extensive investigation of the fluorescence of xanthene dyes I did not observe any quantum yields apparently higher than 100% when the value for fluorescein was taken as 90%, although the ceiling of 100% was reached by some derivatives, e.g. rhodamine 101. I believe that this lends further support to the accuracy of the value 90% for fluorescein. Whereas the quantum yield of this dye seems to be most reliable, its use as a standard solution is hampered by poor chemical stability. Other compounds proposed for this purpose also have more or less serious drawbacks. Rhodamine 6G perchlorate seems to be superior to all the compounds proposed so far for the wavelength region between 500 and 600 nm.

These sketchy remarks on a seemingly simple, yet very difficult problem are intended to warn the reader to use caution when he encounters fluorescence quantum yields. I hasten to point out that, with a little experience, relative quantum yields can be estimated quite well when the solutions are compared visually under an intense light source.

4.3. Intersystem Crossing $S_1 \rightarrow T_1$

Besides the nonradiative decay directly to the ground state, a molecule excited to the state S_1 may enter the system of triplet states and relax to the lowest level T_1. As was pointed out earlier, this is undesirable in a laser dye for various reasons. In this section some structural features are discussed that influence the rate of intersystem crossing in organic dyes. For additional information the reader may turn to [FÖRSTER, 1951 (p. 261); WEHRY, 1967; PARKER, 1968 (p. 303); BIRKS, 1970 (p. 193)].

4.3.1. Dependence on π-Electron Distribution

The intersystem crossing from the singlet to the triplet manifold, being a forbidden process, is comparatively slow considering the small energy gap between S_1 and T_1. Its rate varies, however, in a peculiar fashion with the molecular structure of the dye. Whereas quantum-mechanical concepts have not been applied yet to the classes of dyes which concern us in connection with dye lasers, a simple rule has been found that explains many experimental observations (DREXHAGE, 1972a). This rule states that in a dye where the π-electrons of the chromophore can make a loop when oscillating between the end groups, the triplet yield will be higher than in a related compound where this loop is blocked. It may be said that the circulating electrons create an orbital magnetic moment which couples with the spin of the electrons. This increased spin-orbit coupling then enhances the rate of intersystem crossing, thus giving rise to a higher triplet yield [2].

The above rule may be illustrated by the class of rhodamine and acridine dyes, whose π-electron distribution can be described in terms of mesomeric resonance structures in the following way [3]:

---

[2] This explanation will probably horrify a theoretician. The justification for the loop rule stems mainly from the fact that it is in agreement with experimental observations.

[3] A number of additional mesomeric forms that are of lesser importance are left out here for the sake of clarity.

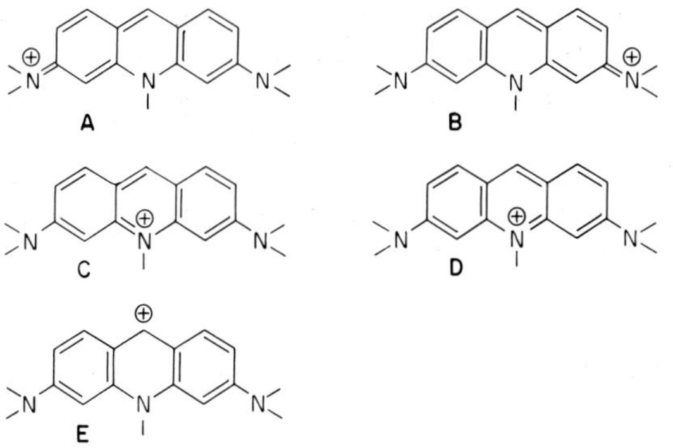

While none of these mesomeric structures alone depicts the real π-electron distribution in the chromophore, they all contribute according to their relative energies. Particular attention must be paid to structures C and D because they symbolize the path the electrons follow along the nitrogen bridge. In the acridine dyes depicted here, structures C and D contain an ammonium configuration, as do structures A and B, so that all of them have approximately the same importance. In the case of rhodamine dyes, however, the central nitrogen atom is replaced by oxygen. Therefore structures C and D now involve an oxonium configuration, which renders them energetically less favored than structures A and B, and they do not contribute as much to the real π-electron distribution as they do in the case of the acridine dyes. Hence, according to the above rule, a lower triplet yield is expected in rhodamine than in acridine dyes. This is confirmed experimentally, as triplet yields of 1% and 10% have been found for rhodamine 6G (WEBB et al., 1970) and acridine orange (SOEP et al., 1972), respectively.

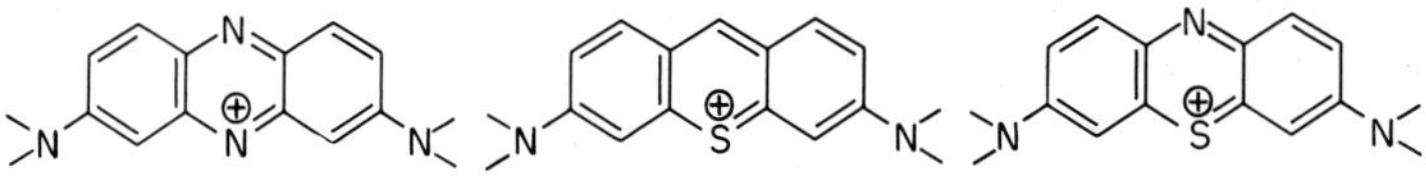

safranin   thiopyronin   thiazine

The loop rule has proven very useful in the design of new laser dyes (DREXHAGE, 1972a, 1973a). It predicts, for instance, a high triplet yield in safranin, thiopyronin, and thiazine dyes, in full agreement with experimental observations (DREXHAGE, 1971). For fluorescein, it predicts a low triplet yield in alkaline solution, yet a high triplet yield in acidic solution, where structures A, B, C, and D all contain the oxonium configuration. These predictions are also borne out by experiments (DREXHAGE, 1971; SOEP et al., 1972). Furthermore, the rule predicts relatively high triplet yields in dyes with a ringlike chromophore, as is the case, e.g. with the chlorophylls and phthalocyanines. On the other hand, a very low triplet yield is expected in oxazine dyes and in compounds that contain a tetrahedral carbon atom in place of the oxygen bridge, and thus do not allow any π-electrons to pass over the bridge (see Section 4.8).

fluorescein, basic   fluorescein, acidic

oxazine   carbazine

4.3.2. Heavy-Atom Substituents

Above we have discussed an intersystem crossing mechanism whose magnitude depends essentially on the π-electron distribution in the chromophore. In the examples given, the molecules were composed exclusively of light atoms, i.e. hydrogen and elements that appear in the first row of the periodic table. However, the intersystem crossing rate, intrinsic to the chromophore, can be greatly enhanced if the dye is substituted with heavier elements, which increase the spin-orbit coupling [TURRO, 1965 (p. 50); WEHRY, 1967; BIRKS, 1970 (p. 208)]. This undesirable effect for a laser dye can be demonstrated on the fluorescein derivatives eosin and erythrosin. The triplet yield of eosin in an alkaline solution was found to be 76%, which is to be compared with the value 3% for fluorescein (SOEP et al., 1972). While substitution with chlorine has very little effect on the intersystem crossing rate of fluorescein, the

eosin   erythrosin   dithiofluorescein

replacement of the oxygen atoms in 3- and 6-position by sulfur, a direct neighbor of chlorine in the periodic table, yields a dye (dithiofluorescein, absorption maximum in alkaline ethanol 635 nm) that is absolutely nonfluorescent (DREXHAGE, 1972b). Obviously, in any planned synthesis of efficient laser dyes, heavy-atom substituents are to be avoided.

4.3.3. Determination of Triplet Yields

Several techniques have been developed to determine the quantum yield of triplet formation from the excited singlet level S_1. However, only a very small number of triplet yields have been reported, illustrating the experimental difficulties encountered in such measurements. For a review the interested reader is referred to [PARKER, 1968 (p. 303); BIRKS, 1970 (p. 193); SOEP et al., 1972]. While all the methods for the measurement of triplet yields published so far require rather elaborate equipment and are very time consuming, a reasonably good estimate of this quantity can be obtained in a few minutes with a new technique which requires only test tubes, an ultraviolet lamp, and a few organic solvents (DREXHAGE, 1971). As this method is particularly convenient in the study of laser dyes, it is described here briefly.

The technique is based on the well-known enhancement of the intersystem crossing rate k_{ST} by solvents containing heavy atoms (see Subsection 4.5.3). It was found that solvents like iodomethane or iodobenzene increase the intrinsic value of k_{ST} in first approximation by a constant factor independent of the magnitude of k_{ST}. Hence in the case of a dye in which the triplet yield is small ($k_{ST} \ll$ rate of fluorescence emission k_{fl}), the enhancement of k_{ST} by a factor of, say, 10^3 will not change the fluorescence efficiency appreciably. However, in the case of a dye where the values of k_{ST} and k_{fl} are comparable, the same enhancement of k_{ST} will cause a reduction of the fluorescence efficiency by three orders of magnitude. Thus, for an approximate determination of the triplet yield, one simply adds an equal amount of iodomethane[4] to the ethanolic dye solution and observes the concomitant reduction in fluorescence intensity. The results obtained with this technique were in agreement with published triplet-yield data in all cases studied, including, among others, aromatic compounds, cyanine dyes, and typical laser dyes like rhodamines. For example, the fluorescence of acriflavine is completely quenched on addition of iodomethane, whereas that of rhodamine 6G remains almost unchanged, demonstrating the large difference in the intersystem crossing rates (see Subsection 4.3.1). Since it is

---

[4] Some compounds are alkylated by iodomethane. In such cases the less reactive iodobenzene is a suitable heavy-atom solvent.

independent of the fate of the triplet-state molecules after their production, this is a most direct technique which enables the determination of an intersystem crossing, even if no triplet molecules can be detected. It is applicable to such fugitive compounds as excimers and exciplexes (see Subsection 4.5.5).

4.4. Light Absorption in the States S_1 and T_1

As with molecules in the ground state S_0, there is a well-defined absorption spectrum associated with molecules that are in the excited states S_1 or T_1. Unfortunately, these spectra are very difficult to measure and accordingly very few data on laser dyes or related compounds have been reported. This is the case particularly with the S_1-absorption, which, owing to the short lifetime of this state, can usually be measured only by laser techniques. While some data are available on aromatic compounds (NOVAK and WINDSOR, 1968; NAKATO et al., 1968; BONNEAU et al., 1968), cyanine (MÜLLER, 1968), and phthalocyanine dyes (MÜLLER and PFLÜGER, 1968), the S_1-absorption spectra of the widely used laser dyes are not known. That this absorption gives rise to appreciable losses when a coumarin laser is pumped at high power densities was suggested recently (WIEDER, 1972).

The absorption spectrum of molecules in the lowest triplet state T_1 is more easily accessible if the compound is embedded in a solid matrix, where the triplet lifetime is very long (in some cases up to several seconds). If the solid solution is excited with a highly intense light source, an appreciable population of the triplet level is achieved, and the spectrum can be determined without much difficulty, in many cases simply with a commercial spectrometer. With such a technique the T_1-absorption spectra of several acridine dyes and of fluorescein in acidic solution have been carefully determined (ZANKER and MIETHKE, 1957a, 1957b; NOUCHI, 1969). Similarly, the T_1-absorption of the laser dyes rhodamine B and rhodamine 6G is known for the red region of the spectrum (BUETTNER et al., 1969; MORROW and QUINN, 1973). It was found that, in the region where these dyes lase, the molar decadic extinction coefficient of the triplet state has a value of $\approx 1.5 \times 10^4\,l$ $\text{mole}^{-1}\,\text{cm}^{-1}$. Recently, the T_1-absorption of 7-diethylamino-4-methylcoumarin has been determined in the region from 400—650 nm (MORROW and QUINN, 1973). In the lasing region of this compound (450—500 nm) the extinction coefficient was found to have a value of $\approx 0.3 \times 10^4\,l\,\text{mole}^{-1}\,\text{cm}^{-1}$. In addition, the triplet absorption of other compounds of interest for dye lasers is given in this paper.

4.5. Environmental Effects

In Sections 4.2 and 4.3 the nonradiative processes in dye molecules have been related to their molecular structure. In the discussion given there, little attention was paid to the environment of the dye molecules, i.e. the solvent and other solute molecules. While such a simplification is often practical, there are also many instances where the surroundings of the dye molecules affect the rates of nonradiative processes to a degree that cannot be neglected.

4.5.1. Fluorescence Quenching by Charge Transfer Interactions

It has been known for a long time that the fluorescence of dyes is quenched by certain anions [PRINGSHEIM, 1949 (p. 322); FÖRSTER, 1951 (p. 181)]. The quenching efficiency depends strongly on the chemical nature of the anion. The quenching ability, which is very strong in the case of iodide (I^-), decreases in the order: Iodide (I^-), thiocyanate (SCN^-), bromide (Br^-), chloride (Cl^-), perchlorate (ClO_4^-). While the mechanism of the phenomenon has not yet been elucidated in detail, this succession suggests that the excited state of the dye is quenched by a charge-transfer interaction. Since this effect is undesirable in lasing solutions, the use of perchlorate as the anion is often preferable to the common chloride, as was demonstrated in the acidic umbelliferone laser (BERGMAN et al., 1972). Because in many laser dyes the chromophore carries a positive charge, these dyes invariably have an anion, commonly chloride or iodide. Whether the fluorescence efficiency is affected by the anion accompanying the chromophore depends on the concentration and the polarity of the solvent. It was found that the fluorescence efficiencies of rhodamine 6G iodide and perchlorate in 10^{-4} molar solutions in ethanol were identical and very high (DREXHAGE, 1972a, 1973a), indicating no quenching by the anions in either case. However, the fluorescence of rhodamine 6G iodide at the same concentration in the nonpolar solvent chloroform was almost completely quenched, whereas the perchlorate was as efficient as in ethanol. Apparently, the dye salts are fully dissociated in the polar solvent ethanol, but practically undissociated in chloroform. Hence, in ethanol, the quenching anions do not have sufficient time to reach the excited dye molecules during their lifetime, whereas, in chloroform, they are immediately available for a reaction. It is therefore advisable to use the dye perchlorate in lasing solutions of low polarity and when high concentrations are required. In addition to the anions mentioned here, a number of other compounds quench the fluorescence by a similar mechanism [PRINGSHEIM, 1949 (p. 322); FÖRSTER, 1951 (p. 181); LEONHARDT and WELLER, 1962].

4.5.2. Quenching by Energy Transfer

Another mechanism by which excited states, singlet as well as triplet, are quenched externally can operate if the quenching molecule has a level of energy equal to or lower than that of the state to be quenched. Under favorable conditions such energy transfer can occur over distances up to about 10 nm [Förster, 1951 (p. 83), 1959; Kellogg, 1970; Birks, 1970 (p. 518)]. In liquid solution, where the reactants can approach each other very closely, energy-transfer processes are very efficient, provided the diffusion time is shorter than the lifetime of the excited state. The main interest in this kind of process in connection with dye lasers arises from its application to the quenching of undesired triplet states. Most widely utilized for this purpose is the ubiquitous molecular oxygen, which is available in sufficient concentration in laser solutions that are exposed to the atmosphere (Snavely and Schäfer, 1969; Marling et al., 1970a). It is not clear yet whether the triplet quenching of oxygen is due to its low-lying excited singlet states or to its paramagnetic properties [Wehry, 1967 (p. 91); Becker, 1969 (p. 230); Birks, 1970 (p. 492)]. However, the quenching of dye triplets by cyclooctatetraene and cycloheptatriene (Pappalardo et al., 1970a) must certainly be attributed to the low-lying triplet levels of these molecules. Several other compounds have been reported to quench triplets of laser dyes (Marling et al., 1970b, 1971). However, in some of these cases it is doubtful whether the quenching action was caused by the compound itself or by some impurity.

4.5.3. External Heavy-Atom Effect

In Subsection 4.3.2. we discussed how substitution with heavier elements enhances the rate of intersystem crossing in organic compounds and thus increases the triplet yield following optical excitation. It is remarkable that the same effect occurs if an organic compound is merely dissolved in a solvent that contains heavy-atom substituents, e.g. iodomethane or iodobenzene. For more information on such phenomena the reader is referred to the literature [Turro, 1965 (p. 57); Wehry, 1967 (p. 86); Birks, 1970 (p. 208)]. Here it may suffice to mention that heavy-atom solvents are not well suited for lasing solutions owing to the increased triplet build-up. How such solvents can be used for the determination of triplet yields is described in Subsection 4.3.3.

4.5.4. Aggregation of Dye Molecules

Since water is a highly desirable solvent for laser dyes, their behavior in this medium has attracted the attention of many workers in this field. It has long been known that organic dyes in aqueous solution have a tendency to form dimers and higher aggregates which make themselves known through a distinctly different absorption spectrum. The dimers usually have a strong absorption band at shorter wavelengths than the monomers and often an additional weaker band at the long-wavelength side of the monomer band [FÖRSTER, 1951 (p. 254); FÖRSTER and KÖNIG, 1957; ROHATGI and MUKHOPADHYAY, 1971; SELWYN and STEINFELD, 1972]. Furthermore, they generally are only weakly fluorescent or not at all. The equilibrium between monomers and dimers shifts to the side of the latter with increasing dye concentration and with decreasing temperature. At ambient temperature and 10^{-4} molar concentration, the dimerization of dyes like rhodamine B and rhodamine 6G is severe enough to prevent laser action. Not only is part of the pump light absorbed by the nonfluorescent dimers, but the dimers also increase the cavity losses owing to their long-wavelength absorption band, which is in the same region as the fluorescence of the monomers.

A number of factors have been suggested as responsible for the aggregation of organic dyes. It has been assumed that an attractive dispersion force between the highly polarizable dye chromophores plays an important role, while the high dielectric constant of water reduces the Coulombic repulsion between the identically charged molecules (RABINOWITCH and EPSTEIN, 1941). However, it was found that the dimerization tendency was much less pronounced when the solvent was formamide, which has an even higher dielectric constant than water (ARVAN and ZAITSEVA, 1961). It has also been suggested that hydrogen bonding between the dye molecules (LEVSHIN and GORSHKOV, 1960; ROHATGI and SINGHAL, 1966), and an interaction with the accompanying anions (LAMM and NEVILLE, 1965; LARSSON and NORDÉN, 1970) may be responsible for the dimerization in certain cases. A systematic study of the class of rhodamine dyes has shown that the dimerization tendency increases with the number and size of alkyl substituents (DREXHAGE and REYNOLDS, 1972). In agreement with earlier suggestions, this indicates a tendency on the part of the hydrophobic dye molecules to shed water in much the same way that organic compounds like benzene avoid water, so that they become insoluble. While aggregation in water is common with most dyes, it usually does not occur in *organic* solvents, even at very high concentration. I do not agree with the claim in a recent report that aggregation of xanthene dyes occurs in ethanol and other organic solvents (SELWYN and STEINFELD, 1972). The dependence of the absorption spectra on concentration and temperature observed by these authors is more probably due to equilibria between different monomeric forms of the dyes (see Section 4.7) (DREXHAGE, 1972a, 1973a).

It is possible to suppress the aggregation of dyes in aqueous solution by the addition of organic compounds. While any compound that is soluble in water will work to a certain degree (CROZET and MEYER, 1970), there are some materials that are exceptionally efficient in this respect, e.g. N,N-dimethyldodecylamine-N-oxide[5] (TUCCIO and STROME, 1972), hexafluoroisopropanol (DREXHAGE, 1972a, 1973a), and N,N-dipropylacetamide (DREXHAGE, 1973a). These compounds can also be used to solubilize dyes that are insoluble in water (TUCCIO et al., 1973). Presumably these additives form a cage around the hydrophobic dye molecules and thus shield them from each other and from the water. In agreement with such a cluster formation, it was observed that the rotational relaxation time of the dye molecules is lengthened in such solutions (SIEGMAN et al., 1972).

4.5.5. Excited State Reactions

Although no interaction between ground-state molecules of organic dyes is evident in organic solvents, there is usually a strong interaction between excited molecules and those in the ground state, and this shows up at high concentrations. While the absorption spectrum of, e.g. rhodamine 6G in ethanol is unchanged even at concentrations as high as 10^{-2} molar, the fluorescence at such concentrations is strongly quenched owing to collisions of the excited dye molecules with those in the ground state [FÖRSTER, 1951 (p.230); BARANOVA, 1965]. It is not known if in cases like this dimeric dye molecules exist for a short time; however, if they do, they are nonfluorescent. Although the fluorescence of the majority of organic compounds is quenched at high concentrations by this mechanism, it has been found that with pyrene and some other compounds a new fluorescence band appears when the concentration is increased [PARKER, 1968 (p.344); FÖRSTER, 1969; BIRKS, 1970 (p.301)]. This new band is due to dimers that exist only in the excited state (excimers). Following the emission of a photon, they immediately dissociate into ground-state monomers.

Similarly, an excited molecule may react with a molecule of a different compound (solvent or other solute) to form an excited complex (exciplex) which, on radiative de-excitation, decomposes immediately into the components [KNIBBE et al., 1968; BIRKS, 1970 (p.403)]. Since the ground state of excimers and exciplexes is unstable, these species are ideal lasing compounds, provided the fluorescence efficiency is high and no disturbing triplet effects occur (SCHÄFER, 1968, 1970). Because some

---

[5] An aqueous solution of this compound is on the market under the tradename Ammonyx-*LO*.

compounds become more basic or acidic on optical excitation, they may pick up a proton from the solution or lose one to the solution (protolysis) [WELLER, 1958; PARKER, 1968 (p. 328); BECKER, 1969 (p. 239)]. If the new forms are fluorescent, they have the same advantage as exciplexes (SCHÄFER, 1968, 1970; SRINIVASAN, 1969; SHANK et al., 1970a). The acid-base equilibria of some coumarin derivatives are discussed in Subsection 4.6.2.

4.6. Coumarin Derivatives

A group of widely used laser dyes emitting in the blue-green region of the spectrum are derived from coumarin by substitution with an amino or hydroxyl group in the 7-position. Since some members of this class rank among the most efficient laser dyes known today, we give here a summary of their pertinent optical properties. The marked change in basicity that occurs on optical excitation causes a shift of the fluorescence to longer wavelengths, a property which may be utilized to achieve a particularly wide tuning range in a dye laser.

4.6.1. Absorption and Fluorescence

The chromophore of such compounds can be described essentially by the mesomeric forms A and B depicted here. In the electronic ground state the π-electron distribution of the molecule closely resembles form A. The comparatively small weight of the polar form B is increased by factors that reduce its energy with respect to form A and a shift of the main absorption band to longer wavelengths occurs. If the weight of structures A and B becomes equal, the coumarin attains the character of a symmetrical cyanine dye and absorption occurs at the longest wavelength possible for this system. The positive charge at the N atom in form B is stabilized, for instance, by electron-donating alkyl groups.

A   **B**

Accordingly, a successive shift of the absorption band to longer wavelengths is found in the series, coumarin 120, coumarin 2, coumarin 1, coumarin 102 (DREXHAGE, 1972a, 1973a). The absorption maximum of these dyes in methanol occurs at 351 nm, 364 nm, 373 nm, and 390 nm, respectively. The more polar mesomeric form B is also stabilized if the dye molecule is surrounded by the molecules of a polar solvent.

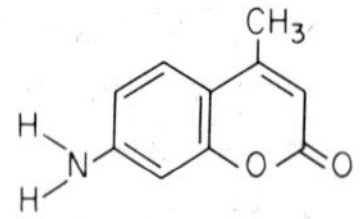

Coumarin 120

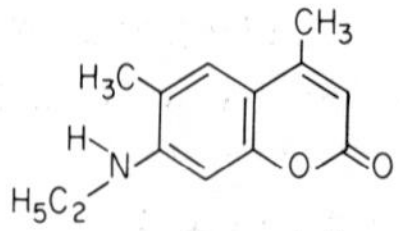

Coumarin 2

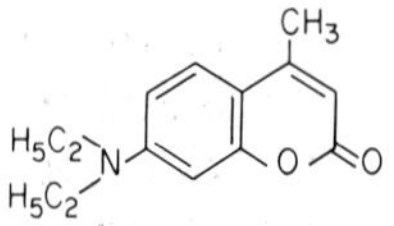

Coumarin 1

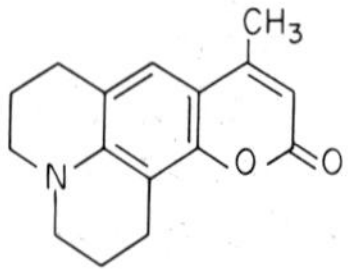

Coumarin 102

Therefore, the absorption maximum should shift to longer wavelengths with increasing polarity of the solvent. The effect can best be seen in the case of coumarin 102 (Table 4.1), where the absorption maximum shifts from 383 nm in N-methyl-pyrrolidinone (NMP) to 418 nm in hexafluoroisopropanol (HFIP). This phenomenon is obscured, however, by a counteracting effect in those coumarins that carry one or two hydrogen atoms at the amino group. Here, polar solvents like water or HFIP exert a specific influence on the amino group, reducing its ability to donate electrons and thereby reducing the weight of structure B. As a consequence, the absorption of coumarin 120 occurs at shorter wavelengths in HFIP than in methanol (Table 4.1). A similar phenomenon is found in the class of xanthene dyes (see Section 4.7).

Table 4.1. Coumarin dyes. λ_{abs} maximum of main absorption band; λ_{las} approximate lasing wavelength (flashlamp-pumped, untuned)

Dye	Solvent[a]	λ_{abs} [nm]	λ_{las} [nm]
Coumarin 120	HFIP	326	
	TFE	338	
	H$_2$O	343	
	MeOH	351	440
Coumarin 2	CH$_2$Cl$_2$	355	
	NMP	362	
	MeOH	364	450
	HFIP	378	460
Coumarin 1	MeOH	373	460
	TFE	388	470
	HFIP	398	480
Coumarin 102	NMP	383	470
	CH$_2$Cl$_2$	386	470
	MeOH	390	480
	TFE	405	500
	HFIP	418	510
Coumarin 30	MeOH	405	510
Coumarin 6	MeOH	455	540

[a] HFIP hexafluoroisopropanol, TFE trifluoroethanol, NMP N-methyl-2-pyrrolidinone.

Whereas in the electronic ground state S_0 of the coumarins the mesomeric structure A is predominant and structure B makes only a minor contribution to the actual π-electron distribution, the opposite is true for the first excited singlet state S_1, in which the polar form B is predominant. Therefore on optical excitation the (static) electric dipole moment increases, and a major rearrangement of the surrounding solvent molecules takes place immediately after excitation. Thus the energy of the excited state is markedly lowered before light emission occurs. This is the reason for the large energy difference between absorption and fluorescence (Stokes shift) in the coumarin derivatives shown in Table 4.1 as compared with, e.g. the xanthene dyes. Without going into detail, we note that the quantum yield of fluorescence has values above 70% in the compounds in Table 4.1.

Coumarin 30

Coumarin 6

Recently it was found that the lasing range covered by coumarin dyes can be extended appreciably if a heterocyclic substituent is introduced into the 3-position (DREXHAGE, 1972a, 1973a). Coumarin 6 in alcoholic solution lases untuned at 540 nm; it is unmatched in efficiency and photochemical stability by other dyes which lase at this wavelength.

An important advantage in flashlamp pumping is the relatively strong absorption of coumarin dyes below 300 nm (Fig. 4.2). It was found that

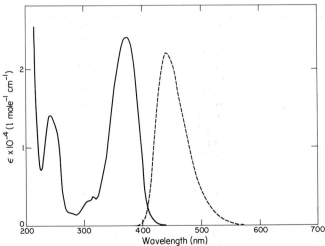

Fig. 4.2. Coumarin 1 in ethanol. ——— Absorption spectrum (ε molar decadic extinction coefficient); – – – quantum spectrum of fluorescence (arbitrary units)

under flashlamp pumping conditions, where the coumarin dyes listed in Table 4.1 lased very efficiently, other dyes having approximately the same long-wavelength absorption and fluorescence efficiency, but very little UV absorption, did not lase at all (DREXHAGE et al., 1972). These same dyes did lase very efficiently, however, when excited with a monochromatic pump source matching their main absorption band. Recently, continuous laser operation with all coumarin derivatives in Table 4.1 was achieved by using aqueous media in which the dyes were rendered soluble by means of suitable additives (Subsection 4.5.4) (TUCCIO et al., 1973).

4.6.2. Acid-Base Equilibria

Owing to the predominance of structure B in the excited state of 7-aminocoumarins, the carbonyl group in 2-position becomes markedly basic following optical excitation and has then an enhanced tendency to pick up a proton from the solution. In the case of coumarin 2, the protonated form created in its excited state exhibits a green fluorescence instead of the blue emission of the unprotonated dye (SRINIVASAN, 1969).

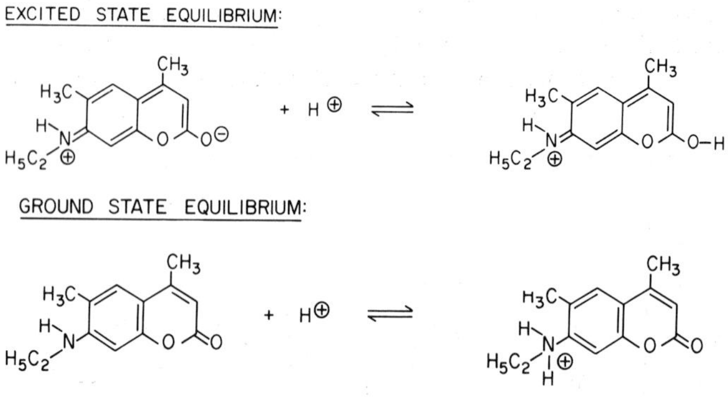

EXCITED STATE EQUILIBRIUM:

GROUND STATE EQUILIBRIUM:

Since the carbonyl group is much less basic in the ground state, the protonated molecules dissociate immediately following the light emission. A dye like coumarin 2 does show some basicity in the ground state, but it is associated with the amino group rather than with the carbonyl group. Therefore, on addition of a few percent of a strong acid (e.g., HCl or $HClO_4$) to an alcoholic solution of coumarin 2, some protonation on the amino group occurs, which causes a reduction of the optical density at 364 nm, as the protonated form does not absorb at this wavelength (DREXHAGE, 1971). Whereas the protonation in the ground state reaches an equilibrium depending on the basicity of the dye and on the $H^{\oplus}$-ion concentration, this is generally not the case for the excited state owing to its short lifetime, on the order of a few nanoseconds. Protonation in the excited state, therefore, is diffusion-controlled. It will increase with the acid concentration, but the undesirable protonation in the ground state increases too, because the equilibrium is dependent on the acid concentration.

While in coumarin 2 protonation in both ground and excited state takes place when its alcoholic solution is acidified with HCl or $HClO_4$, these reactions may be observed separately in other coumarin derivatives (DREXHAGE, 1971). Since the basicity of the amino group is enhanced by electron-donating alkyl substituents, it is highest in a compound carrying two such substituents. Accordingly, it was found that, under the same conditions of acidification as before, coumarin 1 loses its absorption at 373 nm completely owing to protonation of the amino group. On the other hand, in the derivative coumarin 102 no protonation of the amino group is possible for geometric reasons. Hence, on acidification as above, there is no reduction of the optical density and a nearly complete protonation in the excited state is observed. In this case the fluorescence of the protonated form is yellow with a spectrum extending well beyond 600 nm (Fig. 4.3).

Basically, the same protonation-deprotonation reactions as with the 7-aminocoumarins are found for 7-hydroxycoumarins, e.g. 4-methylumbelliferone (coumarin 4) (SHANK et al., 1970a; DIENES et al., 1970). The anion of this compound, which is formed in alkaline solution,

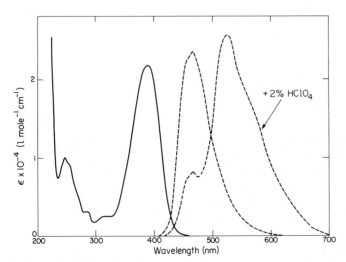

Fig. 4.3. Coumarin 102 in ethanol. ———— Absorption spectrum (ε molar decadic extinction coefficient); – – – quantum spectrum of fluorescence (arbitrary units) in neutral and acidic solution. Quantum yield in acidic solution weaker by a factor of ≈ 2

corresponds closely in its electron distribution to the neutral forms of the 7-aminocoumarins. As with coumarin 2, the anion of coumarin 4 absorbs near 360 nm and fluoresces strongly at ≈ 450 nm. Since in the ground state the mesomeric structure A predominates (see Subsection 4.6.1) and in the excited state structure B, the location of the negative charge and thus of the basicity of the molecule changes as indicated. While this is of no consequence in alkaline solution, it becomes important in neutral or weakly acidic solutions where the compound exists in its neutral form which absorbs near 320 nm (HAMMOND and HUGHES, 1971; NAKASHIMA et al., 1972). Following optical excitation, the proton of the hydroxyl group finds itself suddenly at an oxygen with a positive charge and dissociates away (most easily in

ALKALINE SOLUTION (ANION):

Ground State   Excited State

WEAKLY ACIDIC SOLUTION (NEUTRAL MOLECULE):

Ground State   Excited State
(unstable)

TAUTOMER:

Excited State   Ground State
(unstable)

STRONGLY ACIDIC SOLUTION (CATION):

Ground State   Excited State

solvent mixtures containing water), leaving behind the anion in its excited state. This anion can now either return to the ground state, emitting its characteristic blue fluorescence or, at sufficiently high acid concentration, can pick up a proton from the carbonyl group and form the tautomer of the dye. The latter returns to its ground state, emitting green fluorescence like coumarin 2 in acidic solution. Since the ground state of the tautomer is unstable, the proton will change position, and the stable neutral form of the dye is regenerated. In *strongly* acidic solution the molecule exists as the cation, which absorbs at ≈ 345 nm. Here, no reaction takes place in the excited state, and the cation returns to the ground state, emitting violet fluorescence.

4.7. Xanthene Dyes

Most dye lasers today operate with materials that belong to the class of xanthene dyes. They cover the wavelength region from 500—700 nm and are generally very efficient. Unlike most coumarin derivatives, the xanthene dyes are soluble in water, but tend to form aggregates in this solvent (see Subsection 4.5.4). Fortunately for the development of dye laser applications, derivatives like rhodamine 6G, rhodamine B, and fluorescein have been on the market in good quality (see Section 4.10), and give excellent lasing results even without further purification.

4.7.1. Absorption Spectra

The π-electron distribution in the chromophore of the xanthene dyes can be described approximately by the two identical mesomeric structures, A and B. Several other structures of lesser weight (see Subsection 4.3.1) may be neglected here. Unlike the coumarin dyes,

A   B

forms A and B have the same weight, and thus in the xanthene dyes there is no static dipole moment parallel to the long axis of the molecule in either ground or excited state. For a simplified quantum-mechanical treatment of the main absorption features of these dyes the reader is referred to KUHN [1959 (p. 411)]. The transition moment of the main absorption band, which occurs between 450 nm and 600 nm (Table 4.2), is oriented parallel to the long axis of the molecule. Some of the

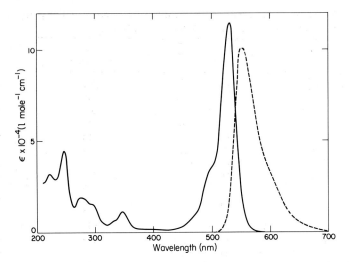

Fig. 4.4. Rhodamine 6 G in ethanol. ——— Absorption spectrum (ε molar decadic extinction coefficient); ——— quantum spectrum of fluorescence (arbitrary units)

transitions at shorter wavelengths (Fig. 4.4), however, are oriented perpendicular to the long axis (JAKOBI and KUHN, 1962).

The position of the long-wavelength band depends markedly on the substituents in 3- and 6-positions of the xanthene nucleus. The absorption maximum in ethanol shifts from 500 nm for the fluorescein dianion to 577 nm for the rhodamine 101 cation (Table 4.2). Since the carboxyphenyl substituent is not part of the chromophore of these dyes

Table 4.2. Rhodamine dyes. λ_{abs} maximum of main absorption band; λ_{las} approximate lasing wavelength (flashlamp-pumped, untuned)

Dye	Solvent[a]	λ_{abs} [nm]	λ_{las} [nm]
Rhodamine 110	HFIP	487	540
	TFE	490	550
	EtOH, basic	501	560
	EtOH, acidic	510	570
	DMSO	518	575
Rhodamine 19	EtOH, basic	518	575
	EtOH, acidic	528	585
Rhodamine 6 G	HFIP	514	570
	TFE	516	575
	EtOH	530	590
	DPA	537	595
	DMSO	540	600
Rhodamine B	EtOH, basic	543	610
	EtOH, acidic	554	620
Rhodamine 3 B	HFIP	550	610
	TFE	550	610
	EtOH	555	620
	DMSO	566	630
Rhodamine 101	HFIP	572	625
	TFE	570	625
	EtOH, basic	564	630
	EtOH, acidic	577	640
	DMSO	586	650

[a] HFIP hexafluoroisopropanol, TFE trifluoroethanol, DMSO dimethyl sulfoxide, DPA N,N-dipropylacetamide.

(see Subsection 4.2.1), it has only a minor influence on the absorption spectrum. Hence the dyes pyronin 20 and pyronin B absorb in ethanol at 527 nm and 552 nm, respectively, i.e. within 2 nm of the corresponding rhodamine cations (Table 4.2). The ultraviolet absorption of the

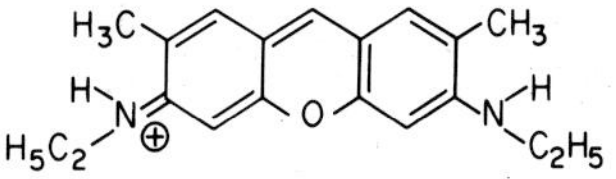

pyronin 20

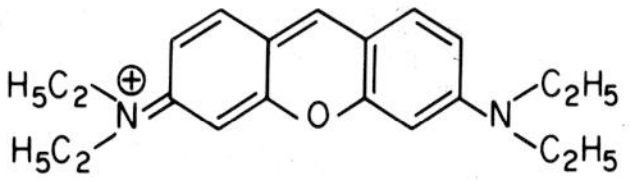

pyronin B

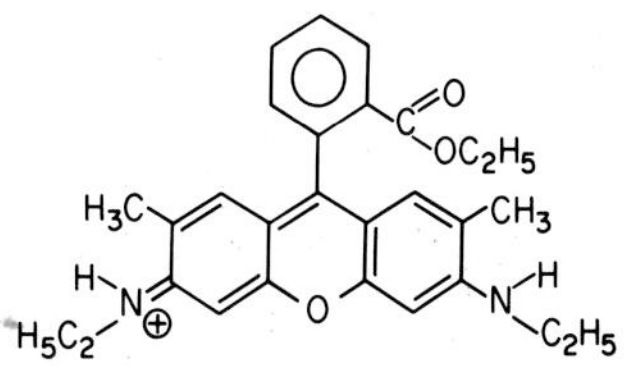

rhodamine 6G

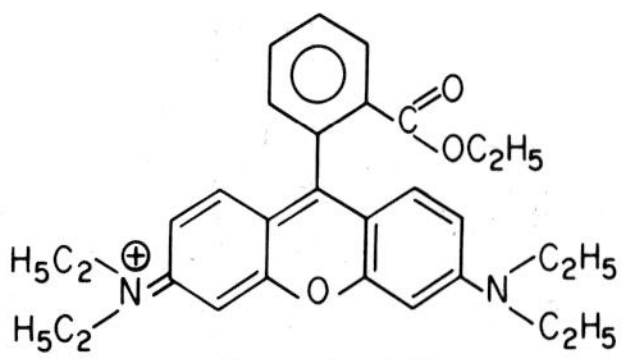

rhodamine 3B

rhodamines is slightly stronger than that of the pyronins. The esters of the rhodamines have absorption spectra practically identical with those of the free acids (Table 4.2). The methyl substituents commonly found in rhodamine 6 G[6] have no influence on either absorption or fluorescence of this dye (Drexhage and Reynolds, 1972). As can be seen from Table 4.2, the absorption maximum of the rhodamines is surprisingly dependent on the solvent, in particular with those dyes whose amino groups are not fully alkylated, e.g. rhodamine 110 and rhodamine 6 G (Drexhage, 1972a, 1973a).

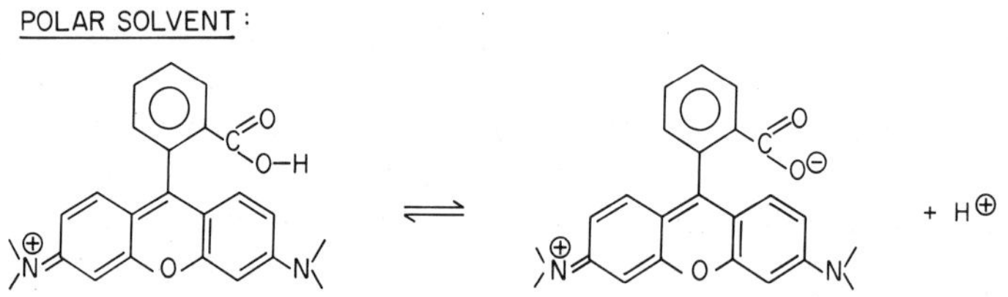

POLAR SOLVENT:

Rhodamine dyes that carry a free (nonesterified) COOH group can exist in several forms. In polar solvents like ethanol or methanol the carboxyl group participates in a typical acid-base equilibrium. The dissociation is enhanced by dilution or, most easily, by adding a small amount of a base. It can be followed spectroscopically, since absorption and fluorescence of the zwitterionic form of the dye are shifted to shorter wavelengths. While this shift amounts to only 3 nm in water (Ramette and Sandell, 1956), it is about 10 nm in ethanol (Table 4.2) (Drexhage, 1972a, 1973a). If a solution of such a dye is prepared in a neutral solvent, it will usually contain both forms of the dye. However, on addition of a small amount of either acid or base, a solution containing essentially only one of the two forms is obtained. In nonpolar solvents, e.g. acetone, the zwitterionic form is not stable.

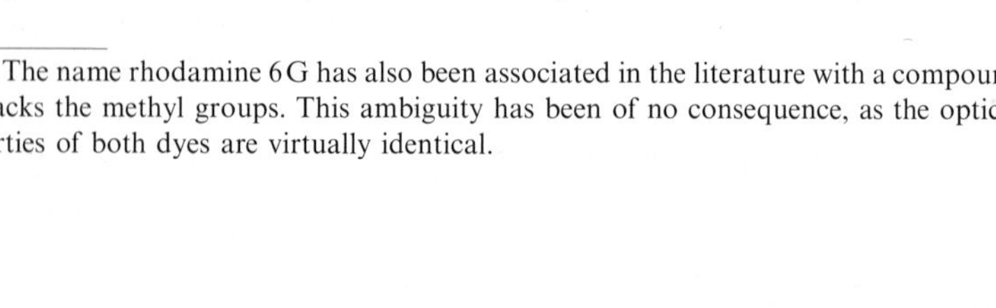

NONPOLAR SOLVENT:

---

[6] The name rhodamine 6 G has also been associated in the literature with a compound that lacks the methyl groups. This ambiguity has been of no consequence, as the optical properties of both dyes are virtually identical.

Instead, an inner lactone is formed in a reversible reaction; this is colorless because the π-electron system of the dye chromophore is interrupted. It may be mentioned that the pyronin dyes also become colorless on addition of a base. However, this reaction is generally not reversible, since in the presence of molecular oxygen a xanthone derivative is rapidly formed, indicated by its strong blue fluorescence.

4.7.2. Fluorescence Properties

The fluorescence spectra of xanthene dyes closely resemble the mirror image of the long-wavelength absorption band (Fig. 4.4). The fluorescence maximum is shifted by about 10 nm in the pyronins and by about 20 nm in the rhodamines with respect to the absorption maximum. It has been shown that the fluorescence quantum yield of rhodamine B in glycerol and in ethylene glycol is independent of the excitation wavelength down to 250 nm (MELHUISH, 1962; YGUERABIDE, 1968). It is reasonable to assume that this is also true for other xanthene dyes. How the fluorescence efficiency depends on the character of the end groups of the chromophore has been discussed in Section 4.2. I do not agree with earlier interpretations (VIKTOROVA and GOFMAN, 1965) which were derived on the assumption that the fluorescence efficiency of the rhodamine dyes would increase with the energy difference between S_1 and S_0. Such an interpretation cannot be upheld, since the new compound rhodamine 101 has the smallest energy difference between S_1 and S_0 of all rhodamines discussed here and yet a fluorescence quantum yield of virtually 100% (DREXHAGE, 1972a, 1973a).

In those xanthene dyes that contain mobile dialkylamino substituents the fluorescence efficiency is influenced by a variety of factors. As already mentioned in Section 4.2, the temperature and viscosity of the solvent affect this important property. It is particularly interesting also that certain low-viscosity solvents appreciably increase the fluorescence efficiency of these dyes above the value found in ethanol (Table 4.3) (DREXHAGE, 1972a, 1973a). The fluorinated alcohols, e.g. trifluoroethanol and hexafluoroisopropanol, provide the additional advantage of disaggregating these dyes in aqueous solutions. Furthermore, it was found that the fluorescence quantum yield of rhodamine B has a value of 40% in acidic ethanol, but of 60% in basic ethanol, whereas the fluorescence efficiencies of rhodamine 110, rhodamine 19, and rhodamine 101 are independent of the acidity (DREXHAGE, 1972b). Apparently a negative charge at the carboxyl group of rhodamine B increases the π-electron density near the remote amino groups sufficiently to cause a reduction in their mobility. This is a

Table 4.3. Fluorescence quantum yield of rhodamine 3 B perchlorate in various solvents at room temperature

Solvent	Qu. Yield [%]
Acetonitrile	27
Acetone	30
Ethanol	42
Chloroform	50
Trifluoroethanol	72
Benzonitrile	75
Dibromomethane	82
Benzaldehyde	82
Benzyl alcohol	87
Hexafluoroacetone hydrate	87
Hexafluoroisopropanol	87
Dichloromethane	91
Glycerol	99

particularly intriguing example of the subtle interplay of the various factors determining the fluorescence efficiency of dyes. It demonstrates how difficult it may be in some cases to unravel the relation between structure and nonradiative processes in organic molecules.

4.8. Newer Classes of Efficient Laser Dyes

In the earlier sections we have outlined a number of structural features that are to be met by efficient laser dyes. On the basis of such principles, novel classes of useful laser dyes can be predicted and prepared by planned synthesis. Here we discuss several newer classes of dyes that have been found very useful in lasers emitting in the red and near infrared.

4.8.1. Oxazine Dyes

If the central $= CH-$ group of a pyronin dye is formally replaced by $= N-$, a compound is obtained whose absorption is shifted by about 100 nm to longer wavelengths [KUHN, 1959 (p.411)]. Such an oxazine or phenoxazine dye is planar and rigid like its xanthene relative. Furthermore, the loop rule (see Subsection 4.3.1) predicts a very low triplet yield, as was confirmed experimentally (DREXHAGE, 1971, 1972a, 1973a). The position of the absorption maximum depends on the end groups of the chromophore in the same fashion as in the xanthene dyes (Table 4.4). The fluorescence maximum shows a Stokes shift of ≈ 30 nm in ethanol (Fig. 4.5), and it was found that these dyes lase quite efficiently at the wavelengths given in Table 4.4. All oxazine dyes are photochemically much more stable than the pyronins and rhodamines.

resorufin

oxazine 118

oxazine 4

oxazine 1

Table 4.4. Oxazine dyes. λ_{abs} maximum of main absorption band; λ_{las} approximate lasing wavelength (laser-pumped, untuned); solvent ethanol

Dye	λ_{abs} [nm]	λ_{las} [nm]
Resorufin	578	610
Oxazine 118	588	630
Oxazine 4	611	650
Oxazine 1	645	715
Oxazine 9	601	645
Nile Blue	635	690

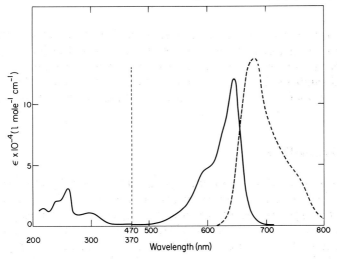

Fig. 4.5. Oxazine 1 in ethanol. ——— Absorption spectrum (ε molar decadic extinction coefficient); --- quantum spectrum of fluorescence (arbitrary units)

   Whereas no triplet problems are encountered in these dyes, it is
comparatively difficult to suppress the processes of internal conversion.
Owing to the smaller energy difference between S_1 and S_0, the influence
of hydrogen vibrations at the end groups is more pronounced than in
xanthene dyes (see Subsection 4.2.2). Hence the use of deuterated alcohol
as a solvent markedly increases the fluorescence efficiency in these dyes
(DREXHAGE, 1972a, 1973a). As in the case of xanthene derivatives, dyes
such as oxazine 1 have a reduced fluorescence efficiency in ethanol owing
to the mobility of the dialkylamino substituents. However, the
fluorescence efficiency of oxazine 1 perchlorate is high with the solvents
dichloromethane, 1,2-dichlorobenzene, or α,α,α-trifluorotoluene. In
such a solvent the dye lases, flashlamp pumped, near 740 nm with
threshold values comparable to that of rhodamine 6G (DREXHAGE,
1972a, 1973a). Fluorinated alcohols are not useful solvents, since they
interact with the central nitrogen atom and so reduce the fluorescence
efficiency of oxazine dyes.
   Some dyes of this class have a structure modified by an additional
benzo group (oxazine 9, nile blue). These have been reported to be

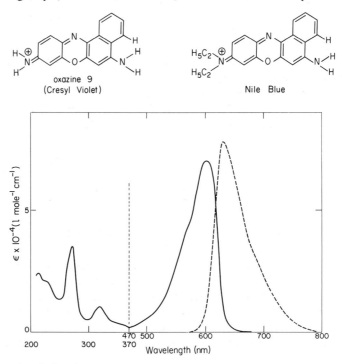

Fig. 4.6. Cresyl Violet in ethanol. ———— Absorption spectrum (ε molar decadic extinction
coefficient); – – – quantum spectrum of fluorescence (arbitrary units)

efficient laser dyes too (MARLING et al., 1970a; RUNGE, 1971)[7]. The additional benzo group causes only a slight shift of absorption and fluorescence toward longer wavelengths but has a distinct influence on the shape of the absorption spectrum (Fig. 4.6). The extinction at the absorption maximum is reduced, and the spectrum is broadened compared with the spectra of other oxazine dyes (Fig. 4.5). This distortion of the absorption spectrum is caused by a steric interference between the amino group and the hydrogen atom indicated in the structural formulas of these dyes (DREXHAGE, 1971). The amount of twist exerted on the amino group depends on the temperature. Hence the familiar shape of the absorption spectrum is restored at low temperatures (GACOIN and FLAMANT, 1972).

4.8.2. Carbon-Bridged Dyes

The xanthene and oxazine dyes discussed above can be considered as diphenylmethane derivatives which are rigidized by an oxygen bridge. Although such a bridge is a good insulator for π-electrons, it still permits some passage of these electrons (see Subsection 4.3.1). To prevent conjugation along the bridge entirely, a tetrahedral carbon atom may be chosen. Whereas such dyes are in general tedious to synthesize, they are very stable once prepared (DREXHAGE and REYNOLDS, 1972). As examples, the dyes carbopyronin 149 and carbazine 122 can be mentioned here. Their absorption maxima in ethanol are at 605 nm and 610 nm, respectively; i.e., they are shifted toward longer wavelengths as compared with the corresponding oxygen-bridged dyes pyronin B and resorufin.

carbopyronin 149   carbazine 122

The fluorescence of these compounds is very strong but the quantum efficiency is slightly reduced by the factors operating in xanthene and oxazine dyes. In the case of carbopyronin 149 the only nonradiative process encountered is associated with the mobility of the dimethylamino end groups. Its influence is reduced if trifluoroethanol is employed as the solvent. The dye in this solvent lases flashlamp-pumped

---

[7] It is not certain whether all reports on cresyl violet lasers were actually based on the dye with the structure given here (see Section 4.10).

at 650 nm with a threshold and efficiency comparable to those of rhodamine 6G. Carbazine 122 has a similarly high fluorescence efficiency. It lases flashlamp-pumped at 720 nm in basic ethanol and at 740 nm in basic dimethyl sulfoxide. It is also soluble in water containing a small amount of base and lases at 700 nm in this solvent.

4.8.3. Naphthofluorescein and Related Dyes

A new class of laser dyes is closely related to the well-known xanthene dyes in that it is derived from α-naphthols instead of phenols (DREXHAGE and REYNOLDS, 1972). These dyes have a longer chromophore than the xanthene derivatives. Therefore they absorb and fluoresce at appreciably longer wavelengths. For instance, the compound naphthofluorescein 126 absorbs in basic ethanol at 615 nm and lases with good efficiency at 700 nm.

fluorescein   naphthofluorescein 126

dye 141   dye 140

Another way of constructing a long, rigid π-electron system is illustrated by the amidopyrylium dyes 141 and 140 (REYNOLDS and DREXHAGE, 1972). These compounds can be considered as pyrylium dyes made rigid by amide groups. The fluorescence efficiency of these dyes is reduced in most solvents (e.g. ethanol) because they carry at least one mobile dimethylamino end group. As in other cases discussed earlier, this mobility can be virtually eliminated by the use of hexafluoroisopropanol as a solvent. In this solvent, dye 141 and dye 140 have an absorption maximum at 590 nm and 660 nm, respectively, and lase very efficiently at 650 nm and 720 nm, respectively.

4.9. Other Efficient Laser Dyes

It is important to note that the structural principles outlined in this article do not constitute *necessary* requirements for efficient laser dyes. In particular, it is quite possible — and not contradictory to the foregoing discussions — that a dye *lacking* structural rigidity has a high fluorescence efficiency, and thus may be a good laser dye. For instance, it has been found that the compounds *p*-terphenyl, *p*-quaterphenyl, and some scintillator dyes, none of which is rigid, are efficient laser dyes in the near-ultraviolet region of the spectrum (FURUMOTO and CECCON, 1970). The finding that the threshold for laser action is higher than in rhodamine 6G can be ascribed in part to the fact that these dyes cannot utilize as much of the flash energy as rhodamine 6G, whose absorption spectrum covers a wider range of frequencies. More recently, MAEDA and MIYAZOE (1972) found that the nonrigid cyanine dyes are valuable laser materials, in particular, when dimethyl sulfoxide is used as a solvent.

Although it is not our intention to provide a comprehensive review of all efficient laser dyes, a few compounds other than those already discussed can be mentioned here. Besides the coumarins, the compounds acridone (FURUMOTO and CECCON, 1969b) and 9-aminoacridine hydrochloride (GREGG and THOMAS, 1969) have been reported to lase efficiently in the blue region of the spectrum. Some quinoline derivatives, which are closely related to the coumarins, are the most efficient compounds known to lase between 400 nm and 430 nm (SRINIVASAN, 1969). The compounds brilliant sulphaflavine (MARLING et al., 1970a) and 6-acetylaminopyrene-1,3,8-trisulfonic acid in alkaline solution (SCHÄFER, 1968, 1970) are excellent lasing materials in the green and yellow. A number of other valuable laser dyes belonging to various classes have been discovered and can be found in the complete list at the end of this article.

4.10. Purity and Chemical Identity of Dyes

Though they may appear mystical to some, organic dyes are, in principle, well-defined compounds. Unfortunately, the purification of dyes is sometimes rather difficult. However, if one succeeds, one may be rewarded with a beautiful fluorescence of the crystalline material, which before was quenched by tiny amounts of impurities. For instance, highly pure rhodamine 6G chloride exhibits a strong red fluorescence, whereas the commercial material is almost nonfluorescent in the crystalline state

(DREXHAGE, 1971). Even so, compared with other commercially available organic dyes, this compound, like rhodamine B, is unusually pure. The quality of many other commercial dyes, however, is not sufficient for laser applications, and purification then yields a material of improved efficiency; see, e.g. GACOIN and FLAMANT (1972).

It is little known, even among scientists working with these materials, that organic dyes are often sold under false names. A recent study has shown that, out of four commercial samples of pyronin B, three were actually rhodamine B and one rhodamine 6G (HOROBIN and MURGATROYD, 1969). Likewise, all four samples of acridine red investigated by the same authors were incorrectly labeled. A dye of this structure should have an absorption maximum of ≈ 530 nm in ethanol and would have lasing properties almost identical with those of rhodamine 6G. Instead, all commercial samples studied by this author absorb at ≈ 550 nm and have optical properties fairly similar to those of pyronin B (DREXHAGE, 1973a). Another example of an ill-defined dye frequently used in dye-lasers is cresyl violet. It is likely that most lasing samples sold under this name have the structure of oxazine 9, which is given in Subsection 4.8.1. But a different structure has also been suggested [Colour Index, 1971 (p. 5406)].

4.11. Concluding Remarks

It was proposed in this chapter to give a critical review of present knowledge on laser dyes. Emphasis was placed on the general concepts relevant to an understanding of their physicochemical properties, and many details had to be omitted. Likewise, in several instances references to the original work were not given; they can be found, however, in the review articles quoted. Although some progress has been made recently in the understanding of nonradiative deactivation processes, this important topic is still far from understood. Many unexplained observations had to be omitted here for the sake of brevity. Furthermore, it must be stressed that present theory is unable to predict the rates of nonradiative processes in dye molecules with any useful degree of accuracy.

The more or less heuristic principles outlined in this article have proven to be very efficient tools in the planned synthesis of new laser dyes and can be expected to lead to further improvements in the near future. One goal will be the extension of the lasing range to shorter as well as to longer wavelengths. In the ultraviolet region, the lasing efficiency decreases with *increasing* energy difference between the levels

S_1 and S_0 owing to increased absorption of the laser light by molecules in S_1. In the infrared, on the other hand, the lasing efficiency decreases with *decreasing* energy difference between S_1 and S_0 owing to increased internal conversion between S_1 and S_0. Hence, it is no coincidence that the most efficient laser dyes known today emit in the visible region of the spectrum. However, efforts are being made currently to construct molecules with high lasing efficiency down to ≈ 300 nm and up to ≈ 1000 nm. Furthermore, it can be expected that in the near future dyes will become available that lase very efficiently in aqueous solution without the addition of any additives. It also should be possible to synthesize good laser dyes that exhibit much greater photochemical stability than, e.g., rhodamines and most coumarins known today.

Acknowledgement. The critical reading of the manuscript by Dr. G. A. REYNOLDS is gratefully acknowledged.

List of Laser Dyes

Dye	Solvent[a]	Excit.[b]	Lasing Wavelength [nm]	Ref.
p-terphenyl	C₆H₁₂	L	341	ABAKUMOV et al. (1969a)
	EtOH	L	340	ABAKUMOV et al. (1969b)
	DMF	F	341	FURUMOTO and CECCON (1970)
p-quaterphenyl	Dioxane	F	343–356	FURUMOTO and CECCON (1970)
	DMF	F	374	FURUMOTO and CECCON (1970)
	C₆H₅CH₃	L	362–390	MYER et al. (1970)
1,4-diphenyl-1,3-butadiene	C₆H₅CH₃	L	383	ABAKUMOV et al. (1969a)
p-distyrylbenzene	C₆H₅CH₃	L	415	KOTZUBANOV et al. (1968a)
p-bis(o-methylstyryl)benzene, bis-MSB	C₆H₅CH₃	F	424	FURUMOTO and CECCON (1969b)
	EtOH	L	419	DEUTSCH and BASS (1969)
	C₆H₆	L	425	DEUTSCH and BASS (1969)
1-(o-methoxystyryl)-4-styrylbenzene	C₆H₅CH₃	L	425	KOTZUBANOV et al. (1968b)
p-bis(o-methoxystyryl)benzene	C₆H₅CH₃	L	430	KOTZUBANOV et al. (1968a)
1-(m-methoxystyryl)-4-styrylbenzene	C₆H₅CH₃	L	415	KOTZUBANOV et al. (1968b)
1-(p-methoxystyryl)-4-styrylbenzene	C₆H₅CH₃	L	425	KOTZUBANOV et al. (1968b)
1-(o-chlorostyryl)-4-styrylbenzene	C₆H₅CH₃	L	420	KOTZUBANOV et al. (1968b)
1-(p-chlorostyryl)-4-styrylbenzene	C₆H₅CH₃	L	420	KOTZUBANOV et al. (1968b)
p-bis(p-chlorostyryl)benzene	C₆H₅CH₃	L	420	KOTZUBANOV et al. (1968a)
4,4'-diphenylstilbene,	C₆H₅CH₃	L	408	KOTZUBANOV et al. (1968a)
1,2-bis(4-biphenylyl)ethylene	C₆H₆	L	409	DEUTSCH and BASS (1969)
	C₆H₆	F	409	FURUMOTO and CECCON (1970)
	DMF	F	409	FURUMOTO and CECCON (1970)
1-(p-phenylstyryl)-4-styrylbenzene, 1-styryl-4-[2-(4-biphenylyl)vinyl]-benzene	C₆H₅CH₃	L	432	KOTZUBANOV et al. (1968a)
4,4'-(p-phenylenedivinylene)bis[N,N-dimethylaniline]	PMMA	L	478	NABOIKIN et al. (1970)
1,2-di-1-naphthylethylene	C₆H₅CH₃	L	426	KOTZUBANOV et al. (1968a)

1-[2-(2-naphthyl)vinyl]-4-styrylbenzene, 1-styryl-4-[2-(2-naphthyl)vinyl]-benzene	$C_6H_5CH_3$	425	L	Kotzubanov et al. (1968b)
2,5-diphenylfuran, PPF	C_6H_6, EtOH	369–380	L	Broida and Haydon (1970)
	EtOH	365–371	L	Myer et al. (1970)
	DMF	371	F	Furumoto and Ceccon (1970)
	Dioxane	371	F	Furumoto and Ceccon (1970)
1,3-diphenylisobenzofuran	EtOH	484–518	F	Marling et al. (1970a)
2,5-diphenyloxazole, PPO	C_6H_{12}	357	L	Lidholt and Wladimiroff (1970)
	$C_6H_5CH_3$, C_6H_6, dioxane	359–391	L	Broida and Haydon (1970)
	Dioxane	381	F	Furumoto and Ceccon (1970)
2,5-bis(4-biphenylyl)oxazole, BBO	$C_6H_5CH_3$	409	L	Abakumov et al. (1969a)
	C_6H_6	409	L	Deutsch and Bass (1969)
	C_6H_6	410	F	Furumoto and Ceccon (1970)
	DMF	413	L	Lidholt and Wladimiroff (1970)
2-(1-naphthyl)-5-phenyloxazole, α-NPO	$C_6H_5CH_3$	400	L	Abakumov et al. (1969a)
	C_6H_{12}	393	L	Deutsch and Bass (1969)
	EtOH	398	L	Deutsch and Bass (1969)
	EtOH	400	F	Furumoto and Ceccon (1970)
	PMMA	396	L	Naboikin et al. (1970)
	PSt	411	L	Naboikin et al. (1970)
2,2'-p-phenylenebis(5-phenyloxazole), 1,4-bis[2-(5-phenyloxazolyl)]benzene, POPOP	$C_6H_5CH_3$	417	L	Kotzubanov et al. (1968a)
	C_6H_{12}	411	L	Deutsch and Bass (1969)
	EtOH	421	L	Deutsch and Bass (1969)
	$C_6H_5CH_3$	419	F	Furumoto and Ceccon (1969)
	THF	415–430	L	Myer et al. (1970)
	PMMA	415	L	Naboikin et al. (1970)
	PSt	420	L	Naboikin et al. (1970)

[a] CH_2Cl_2 dichloromethane, CH_3CN acetonitrile, C_5H_5N pyridine, $C_5H_{11}OH$ isoamyl alcohol, $C_6H_4Cl_2$ o-dichlorobenzene, $C_6H_5CH_3$ toluene, $C_6H_5CF_3$ α,α,α-trifluorotoluene, $C_6H_5NO_2$ nitrobenzene, C_6H_6 benzene, $C_6H_{10}O$ cyclohexanone, C_6H_{12} cyclohexane, C_7H_{16} n-heptane, DCE dichloroethane, DMF N,N-dimethylformamide, DMSO methyl sulfoxide, DPA N,N-dipropylacetamide, EG ethylene glycol, EtOH ethanol, HFIP hexafluoroisopropanol, H_2O water, H_2SO_4 sulfuric acid, IP isopentane, MCH methylcyclohexane, MeCs methyl cellosolve, MeOD monodeuterated methanol, MeOH methanol, NMP 1-methyl-2-pyrrolidinone, PMMA polymethylmethacrylate, PSt polystyrene, TFE trifluoroethanol, THF tetrahydrofuran.

[b] L laser-pumped, F flashlamp-pumped, C continuous (laser-pumped).

List of Laser Dyes (cont.)

Dye	Solvent[a]	Excit.[b]	Lasing Wavelength [nm]	Ref.
2,2'-p-phenylenebis(4-methyl-5-phenyl-oxazole),	C_6H_{12}	L	423	DEUTSCH and BASS (1969)
	EtOH	L	431	DEUTSCH and BASS (1969)
1,4-bis[2-(4-methyl-5-phenyloxa-zolyl)]benzene, dimethyl-POPOP	C_6H_6, $C_6H_5CH_3$, C_6H_{12}	L	424–441	BROIDA and HAYDON (1970)
2,2'-p-phenylenebis[5-(4-biphenylyl)oxazole], BOPOB	THF	L	428–450	MYER et al. (1970)
2,2'-p-phenylenebis[5-(1-naphthyl)oxazole], α-NOPON	C_6H_6	L	430–455	MYER et al. (1970)
5-phenyl-2-(p-styrylphenyl)oxazole	$C_6H_5CH_3$	L	417	NABOIKIN et al. (1970)
	PMMA	L	414	NABOIKIN et al. (1970)
5-phenyl-2-[p-(p-phenylstyryl)phenyl]oxazole	$C_6H_5CH_3$	L	432	NABOIKIN et al. (1970)
	PMMA	L	428	NABOIKIN et al. (1970)
2-(4-biphenylyl)-5-(1-naphthyl)oxazole, BzNO	$C_6H_5CH_3$	L	413	NABOIKIN et al. (1970)
	PMMA	L	410	NABOIKIN et al. (1970)
5-(4-biphenylyl)-2-[p-(4-phenyl-1,3-butadienyl)phenyl]oxazole	$C_6H_5CH_3$	L	448	NABOIKIN et al. (1970)
	PMMA	L	443	NABOIKIN et al. (1970)
2,2'-[p-phenylenebis(vinylene-p-phe-nylene)]bis[5-(4-biphenylyl)oxazole]	$C_6H_5CH_3$	L	460	NABOIKIN et al. (1970)
	PMMA	L	455	NABOIKIN et al. (1970)
2-[p-[2-(2-naphthyl)vinyl]phenyl]-5-phenyloxazole	$C_6H_5CH_3$	L	428	NABOIKIN et al. (1970)
	PMMA	L	425	NABOIKIN et al. (1970)
2-[p-[2-(9-anthryl)vinyl]phenyl]-5-phenyloxazole	PMMA	L	480	NABOIKIN et al. (1970)
2,2'-vinylenebis[5-phenyloxazole]	$C_6H_5CH_3$	L	447	NABOIKIN et al. (1970)
	PMMA	L	444	NABOIKIN et al. (1970)
2,2'-(2,5-thiophenediyl)bis(5-tert-butyl-benzoxazole), 2,5-bis[5-tert-butylbenzoxazolyl(2)]thiophene, BBOT	C_6H_6	F	437	FURUMOTO and CECCON (1970)

Compound	Solvent	Wavelength		Reference
2,5-diphenyl-1,3,4-oxadiazole, PPD	EtOH	347	L	Abakumov et al. (1969b)
	Dioxane	348	F	Furumoto and Ceccon (1970)
2-(p-methoxyphenyl)-5-phenyl-1,3,4-oxadiazole	EtOH	365	L	Abakumov et al. (1969b)
2,5-bis(p-methoxyphenyl)-1,3,4-oxadiazole	EtOH	359, 372	L	Abakumov et al. (1969b)
2-[p-(dimethylamino)phenyl]-5-phenyl-1,3,4-oxadiazole	$C_6H_5CH_3$	408	L	Naboikin et al. (1970)
2-(p-chlorophenyl)-5-[p-(dimethylamino)phenyl]-1,3,4-oxadiazole	$C_6H_5CH_3$	420	L	Naboikin et al. (1970)
2-(4-biphenylyl)-5-phenyl-1,3,4-oxadiazole,	EtOH	363	F	Furumoto and Ceccon (1970)
	EtOH	355–382	L	Myer et al. (1970)
2-phenyl-5-(4-biphenylyl)-1,3,4-oxadiazole, PBD	$C_6H_5CH_3$	377–415	L	Broida and Haydon (1970)
2-(4-biphenylyl)-5-p-cumenyl-1,3,4-oxadiazole, isopropyl-PBD	C_6H_{12}	361	F	Furumoto and Ceccon (1970)
	EtOH	370	F	Furumoto and Ceccon (1970)
2,5-bis(4-biphenylyl)-1,3,4-oxadiazole, BBD	$C_6H_5CH_3$	363–385	L	Turek and Yardley (1971)
	$C_6H_5CH_3$	374–398	L	Turek and Yardley (1971)
2-(4-biphenylyl)-5-styryl-1,3,4-oxadiazole	$C_6H_5CH_3$	391	L	Abakumov et al. (1969a)
2,5-di-1-naphthyl-1,3,4-oxadiazole, α-NND	$C_6H_5CH_3$	391	L	Abakumov et al. (1969a)
2-(1-naphthyl)-5-styryl-1,3,4-oxadiazole	$C_6H_5CH_3$	399	L	Naboikin et al. (1970)
	PMMA	416	L	Naboikin et al. (1970)
2-(4-biphenylyl)-5-(p-styrylphenyl)-1,3,4-oxadiazole	$C_6H_5CH_3$	403	L	Naboikin et al. (1970)
	PMMA	400	L	Naboikin et al. (1970)
2-phenyl-5-[p-(4-phenyl-1,3-butadienyl)phenyl]-1,3,4-oxadiazole	$C_6H_5CH_3$	424	L	Naboikin et al. (1970)
	PMMA	418	L	Naboikin et al. (1970)
1,5-diphenyl-3-styryl-2-pyrazoline	PMMA	450	L	Naboikin et al. (1970)
3-(p-chlorostyryl)-1,5-diphenyl-2-pyrazoline	PMMA	457	L	Naboikin et al. (1970)
Salicylic acid (sodium salt)	EtOH	395–418	L	Myer et al. (1970)
Aminobenzoic acid	$C_6H_5CH_3$	398–406	L	Myer et al. (1970)
Amino G acid	H_2O	459	F	Maeda and Miyazoe (1972)

List of Laser Dyes (cont.)

Dye	Solvent[a]	Excit.[b]	Lasing Wavelength [nm]	Ref.
Anthracene	Fluorene (crystal)	L	408	KARL (1972)
9-methylanthracene	EtOH-MeOH(glass)	L	414	FERGUSON and MAU (1972a)
9,10-dimethylanthracene	MCH-$C_6H_5CH_3$(glass)	L	432	FERGUSON and MAU (1972a)
9-chloroanthracene	EtOH-MeOH(glass)	L	416	FERGUSON and MAU (1972a)
9-phenylanthracene	EtOH-MeOH(glass)	L	417	FERGUSON and MAU (1972a)
9,10-diphenylanthracene	C_6H_{12}	L	433	HUTH and FARMER (1968)
	EtOH	L	435–450	MYER et al. (1970)
	EtOH-MeOH(glass)	L	430	FERGUSON and MAU (1972a)
2-amino-7-nitrofluorene	$C_6H_4Cl_2$	L	540–585	GRONAU et al. (1972)
3-aminofluoranthene	EtOH	F	548–580	MARLING et al. (1970a)
8-hydroxypyrene-1,3,6-trisulfonic acid (sodium salt)	H_2O(basic)	F	550	DREXHAGE (1971)
8-ethylaminopyrene-1,3,6-trisulfonic acid (sodium salt)	H_2O	L	441	SCHÄFER et al. (1967)
8-acetamidopyrene-1,3,6-trisulfonic acid (sodium salt)	H_2O(neutral)	L, F	441–453	SCHÄFER (1970)
	H_2O(basic)	L, F	566–574	SCHÄFER (1970)
Perylene	DMF	L	472	LIDHOLT and WLADIMIROFF (1970)
	DCE	L	472–476	HAMMOND and HUGHES (1971)
Coronene	MCH-IP(glass)	F	444	KOHLMANNSPERGER (1969)
7-amino-2-hydroxy-4-methylquinoline, carbostyril 124	EtOH	F	413	SRINIVASAN (1969)
7-dimethylamino-2-hydroxy-4-methyl-quinoline, carbostyril 165	EtOH	F	425	SRINIVASAN (1969)
7-hydroxycoumarin, umbelliferone	H_2O(basic)	F	457	SNAVELY and PETERSON (1968)
	EtOH(acidic)	L	405–568	DIENES et al. (1970)
7-hydroxy-4-methylcoumarin, 4-methylumbelliferone	H_2O(basic)	F	454	SNAVELY et al. (1967)
	EtOH(acidic)	L	391–567	SHANK et al. (1970)
coumarin 4	H_2O(basic)	C	460	TUCCIO et al. (1973)
Esculin	H_2O(basic)	F	blue	SOROKIN et al. (1968)

Compound	Solvent		Wavelength	Reference
4,8-dimethyl-7-hydroxycoumarin, 4,8-dimethylumbelliferone	EtOH(basic)	F	455–505	GREGG et al. (1970)
	EtOH(acidic)	L	447–569	HAMMOND and HUGHES (1971)
Benzyl-β-methylumbelliferone	EtOH-H₂O(basic)	F	464–468	GREGG and THOMAS (1969)
4-methylumbelliferone-methylene-iminodiacetic acid	EtOH(basic)	F	459–464	MARLING et al. (1970a)
Calcein blue	EtOH(basic)	F	449–490	MARLING et al. (1970a)
7-hydroxy-6-methoxycoumarin	H₂O(basic)	F	blue	CROZET et al. (1971)
3,4-cyclopentano-7-hydroxycoumarin	H₂O(basic)	F	blue	CROZET et al. (1971)
7-acetoxy-4-methylcoumarin	EtOH(basic)	F	441–486	GREGG et al. (1970)
7-butyroxy-4-methylcoumarin	H₂O-CH₃CN(basic)	F	blue	CROZET et al. (1971)
7-acetoxy-5-allyl-4,8-dimethyl-coumarin	EtOH(basic)	F	458–515	GREGG et al. (1970)
7-amino-4-methylcoumarin, coumarin 120	MeOH	F	440	DREXHAGE (1972a, 1973a)
	H₂O-DPA	C	450	TUCCIO et al. (1973)
4,6-dimethyl-7-methylaminocoumarin, coumarin 9	EtOH	F	443	SRINIVASAN (1969)
	EtOH(acidic)	F	484	SRINIVASAN (1969)
	EtOH(acidic)	L	430–530	GUTFELD et al. (1970)
4,6-dimethyl-7-ethylaminocoumarin, coumarin 2	EtOH	F	446	SRINIVASAN (1969)
	EtOH(acidic)	F	487	SRINIVASAN (1969)
	HFIP	F	460	DREXHAGE (1973a)
	H₂O-DPA	C	460	TUCCIO et al. (1973)
7-diethylamino-4-methylcoumarin, coumarin 1	EtOH	F	460	KAGAN et al. (1968)
	EtOH	F	460	FERRAR (1969b)
	HFIP	F	480	DREXHAGE (1972a, 1973a)
	TFE	F	470	DREXHAGE (1973a)
Coumarin 102	MeOH	F	480	DREXHAGE (1972a, 1973a)
	HFIP	F	510	DREXHAGE (1972a, 1973a)
	NMP	F	470	DREXHAGE (1973a)
	TFE	F	500	DREXHAGE (1973a)
	H₂O-DPA	C	480	TUCCIO et al. (1973)
7-amino-3-phenylcoumarin, coumarin 10	H₂O-DPA	C	480	TUCCIO et al. (1973)
Coumarin 24	MeOH	F	510	DREXHAGE (1972a)
Coumarin 30	MeOH	F	510	DREXHAGE (1973a)
	H₂O-DPA	C	505	TUCCIO et al. (1973)

List of Laser Dyes (cont.)

Dye	Solvent[a]	Excit.[b]	Lasing Wavelength [nm]	Ref.
Coumarin 7	H_2O-DPA	C	525	Tuccio et al. (1973)
Coumarin 6	MeOH	F	540	Drexhage (1972a, 1973a)
	H_2O-DPA	C	540	Tuccio et al. (1973)
2,4,6-triphenylpyrylium fluoborate	MeOH	L	485	Schäfer et al. (1967)
2,4,6-tri-p-tolylpyrylium perchlorate	MeOH	L	492	Schäfer (1970)
2,6-bis(p-methoxyphenyl)-4-phenyl-pyrylium fluoborate	MeOH	L, F	566–573	Schäfer (1970)
9(10H)-acridone	EtOH	L	439	Sorokin et al. (1967)
	EtOH	F	435	Furumoto and Ceccon (1969)
9-aminoacridine hydrochloride	MeOH	F	449–453	Schäfer (1970)
N-methylacridinium perchlorate	EtOH, H_2O(acidic)	F	457–460	Gregg and Thomas (1969)
	MeCs	L	535–553	Hammond and Hughes (1971)
Lucigenin, bis-N-methylacridinium nitrate	H_2O(acidic)	L	600	Stepanov and Rubinov (1968)
Acriflavine, trypaflavine	EtOH	L	510	McFarland (1967)
	EtOH	F	517	Furumoto and Ceccon (1970)
Acridine yellow	EtOH	F	514	Maeda and Miyazoe (1972)
3-aminophthalimide	$C_5H_{11}OH$	L	500	Stepanov and Rubinov (1968)
3-amino-N-methylphthalimide	EtOH	L	519	Neporent and Shilov (1971)
3-amino-6-dimethylamino-N-methyl-phthalimide	C_5H_5N	L	560–588	Sevchenko et al. (1968a)
3,6-bis(methylamino)-N-methyl-phthalimide	C_5H_5N	L	578–605	Sevchenko et al. (1968a)
3-dimethylamino-6-methylamino-N-methylphthalimide	C_5H_5N	L	580–593	Sevchenko et al. (1968a)
	$C_6H_5CH_3$	L	578	Aristov et al. (1970)
	EtOH	L	595	Aristov et al. (1970)
	Glycerol	L	610	Aristov et al. (1970)
	C_7H_{16}-EtOH	L	560–590	Aristov and Kuzin (1971)
3,6-bis(dimethylamino)-N-methyl-phthalimide	C_5H_5N	L	575–590	Sevchenko et al. (1968a)
	C_6H_6	L	577	Aristov et al. (1970)
	$C_6H_{10}O$	L	578	Aristov et al. (1970)
	EtOH	L	584	Aristov et al. (1970)

Dye	Solvent	Wavelength	Type	Reference
3-acetamido-6-amino-N-methyl-phthalimide	C₅H₅N	570–600	L	Sevchenko et al. (1968a)
Brilliant sulphaflavine	EtOH	508–573	F	Marling et al. (1970a)
Fluorescein (sodium salt), uranine	H₂O(basic)	539	L	Schäfer et al. (1967)
	EtOH	green	L	Sorokin et al. (1967)
	H₂O, EtOH	550	F	Sorokin and Lankard (1967)
	EtOH	555	C	Strome and Tuccio (1971)
	H₂O-(AmmonyxLO)	522–570	C	Hercher and Pike (1971b)
6-carboxyfluorescein	EtOH(basic)	539–548	F	Marling et al. (1970a)
Fluorescein isothiocyanate	EtOH(basic)	546	F	Maeda and Miyazoe (1972)
Fluorescein diacetate, diacetylfluorescein	EtOH(basic)	541–571	F	Gregg et al. (1970)
2,7'-dichlorofluorescein	EtOH(basic)	green	F	Sorokin et al. (1968)
	EtOH(basic)	575	C	Strome and Tuccio (1971)
2',4',5',7'-tetrachlorofluorescein	EtOH(basic)	582	C	Strome and Tuccio (1971)
Monobromofluorescein	Glycerol	560	L	Stepanov and Rubinov (1968)
Dibromofluorescein	Glycerol	568	L	Stepanov and Rubinov (1968)
Eosin(sodium salt),	EtOH	yellow	L	Sorokin et al. (1967)
2',4',5',7'-tetrabromofluorescein (sodium salt)	MeOH(basic)	553–557	L	Schäfer (1970)
	DMF	568	L	Lidholt and Wladimiroff (1970)
Rhodamine 110, unsubstituted rhodamine		525–585	C	Tuccio and Strome (1972)
	HFIP	540	F	Drexhage (1972a, 1973a)
	TFE	550	F	Drexhage (1973a)
	EtOH(basic)	560	F	Drexhage (1973a)
	EtOH(acidic)	570	F	Drexhage (1973a)
	DMSO	575	F	Drexhage (1973a)
Rhodamine 19	EtOH(basic)	575	F	Drexhage (1973a)
	EtOH(acidic)	585	F	Drexhage (1973a)
Rhodamine 6G	EtOH	green-orange	L	Sorokin et al. (1967)
	EtOH	585	F	Sorokin and Lankard (1967)
	MeOH	598	F	Schmidt and Schäfer (1968)
	Glycerol	557	L	Kotzubanov et al. (1968a)
	PMMA	601	F	Peterson and Snavely (1968)
	H₂O-(TritonX100)	597	C	Peterson et al. (1970)
	H₂O-HFIP	540–605	C	Tuccio and Strome (1972)

List of Laser Dyes (cont.)

Dye	Solvent[a]	Excit.[b]	Lasing Wavelength [nm]	Ref.
Rhodamine 6G	H_2O-(AmmonyxLO)	C	565–640	Tuccio and Strome (1972)
	H_2O-(AmmonyxLO)	C	590–610	Ippen et al. (1972b)
	HFIP	F	570	Drexhage (1972a, 1973a)
	DMSO	F	600	Drexhage (1972a, 1973a)
	TFE	F	575	Drexhage (1973a)
	DPA	F	595	Drexhage (1973a)
N,N′-bis(β-phenylethyl)rhodamine	EtOH	L	569	Aristov et al. (1970)
Rhodamine B, N,N,N′,N′-tetraethylrhodamine	EtOH	L	608	Schäfer et al. (1967)
	MeOH	F	rot	Schmidt and Schäfer (1967)
	PMMA	F	625	Peterson and Snavely (1968)
	H_2O-HFIP	C	580–655	Tuccio and Strome (1972)
	EtOH	F	605–645	Arthurs et al. (1972)
	EtOH(basic)	F	610	Drexhage (1972a, 1973a)
	EtOH(acidic)	F	620	Drexhage (1972a, 1973a)
Rhodamine 3B	C_5H_{11}OH	L, F	615	Rubinov and Mostovnikov (1968)
	HFIP	F	610	Drexhage (1972a, 1973a)
	EtOH	F	620	Drexhage (1972a, 1973a)
	TFE	F	610	Drexhage (1973a)
	DMSO	F	630	Drexhage (1973a)
N,N′-bispentamethylenerhodamine	EtOH	L	599	Aristov et al. (1970)
Rhodamine 101	EtOH(basic)	F	630	Drexhage (1972a, 1973a)
	EtOH(acidic)	F	640	Drexhage (1972a, 1973a)
	DMSO	F	650	Drexhage (1972a, 1973a)
	HFIP	F	625	Drexhage (1973a)
	TFE	F	625	Drexhage (1973a)
Rhodamine S, rhodamine C	C_5H_{11}OH	L, F	620	Rubinov and Mostovnikov (1968)
	EtOH	L	570	Kotzubanov et al. (1968a)
Lissamine rhodamine B-200	EtOH	F	578–595	Gregg and Thomas (1969)
Xylene red B	EtOH	F	575–645	Marling et al. (1970a)
	EtOH(basic)	F	584–645	Gregg et al. (1970)

Dye	Solvent	λ		Reference
Kiton red S	EtOH	589–642	F	GREGG et al. (1970)
Rhodamine 6Y	EtOH	572	L	ARISTOV et al. (1970)
Acridine red	EtOH	orange	L	SOROKIN et al. (1967)
	EtOH	602	F	SOROKIN and LANKARD (1967)
Pyronin G, pyronin Y	$C_5H_{11}OH$	600	L, F	RUBINOV and MOSTOVNIKOV (1968)
	$C_5H_{11}OH$	585	F	STEPANOV and RUBINOV (1968)
	EtOH(acidic)	590–635	F	MARLING et al. (1970a)
Pyronin B	MeOH	gelb	F	SCHMIDT and SCHÄFER (1967)
	PMMA	576	L	KOTZUBANOV et al. (1968a)
Safranin T	EtOH	615–632	F	GREGG and THOMAS (1969)
	EtOH	610	L	KOTZUBANOV et al. (1968a)
	MeOH	621–625	L	SCHÄFER (1970)
Resorufin	EtOH(basic)	610	L	DREXHAGE (1972a, 1973a)
Resazurin	MeOH, EG		C	RUNGE (1972)
Oxazine 118	EtOH	630	L	DREXHAGE (1972a, 1973a)
Oxazine 4	EtOH	650	L	DREXHAGE (1972a, 1973a)
Oxazine 1	EtOH	715	L	DREXHAGE (1972a, 1973a)
	$C_6H_5CF_3$	730	F	DREXHAGE (1972a)
	$C_6H_4Cl_2$	740	F	DREXHAGE (1972a, 1973a)
	CH_2Cl_2	740	F	DREXHAGE (1972a, 1973a)
Cresyl violet, oxazine 9	EtOH	646–709	F	MARLING et al. (1970a)
	MeOH	650	C	RUNGE (1971)
Nile blue	EtOH	644–704	F	ARTHURS et al. (1972b)
	MeOD	680	F	DREXHAGE (1972a, 1973a)
Nile blue A	EtOH	690	L	DREXHAGE (1972a, 1973a)
	MeOH		C	RUNGE (1971)
Carbopyronin 149	TFE	650	F	DREXHAGE and REYNOLDS (1972)
Carbazine 122	EtOH(basic)	720	F	DREXHAGE and REYNOLDS (1972)
	DMSO(basic)	740	F	DREXHAGE and REYNOLDS (1972)
Thionin	H_2SO_4	850	L	RUBINOV and MOSTOVNIKOV (1967)
Methylene blue	H_2SO_4	835	L	STEPANOV et al. (1967)
Toluidine blue	H_2SO_4	848	L	RUBINOV and MOSTOVNIKOV (1967)
Naphthofluorescein 126	EtOH(basic)	700	F	DREXHAGE and REYNOLDS (1972)
Amidopyrylium dye 141	HFIP	650	F	REYNOLDS and DREXHAGE (1972)
Amidopyrylium dye 140	HFIP	720	F	REYNOLDS and DREXHAGE (1972)

List of Laser Dyes (cont.)

Dye	Solvent[a]	Excit.[b]	Lasing Wavelength [nm]	Ref.
Lachs	Glycerol	L	540	STEPANOV and RUBINOV (1968)
Pinaorthol	EtOH	L	565	STEPANOV and RUBINOV (1968)
Violettrot	$C_5H_{11}OH$	L	620	STEPANOV and RUBINOV (1968)
	$C_5H_{11}OH$	F	610	STEPANOV and RUBINOV (1968)
Isoquinoline red	H_2O	L	620	STEPANOV and RUBINOV (1968)
Rapid-filter gelb	$C_5H_{11}OH$	L	620	STEPANOV and RUBINOV (1968)
	$C_5H_{11}OH$	F	610	STEPANOV and RUBINOV (1968)
Echtblau B, pure blue	Glycerol	L	753	RUBINOV and MOSTOVNIKOV (1967)
Naphthalene green	Glycerol	L	756	RUBINOV and MOSTOVNIKOV (1967)
Rhoduline blue 6G	Glycerol	L	758	RUBINOV and MOSTOVNIKOV (1967)
Brilliant green	Glycerol	L	759	RUBINOV and MOSTOVNIKOV (1967)
Victoria blue	Glycerol	L	809	RUBINOV and MOSTOVNIKOV (1967)
Victoria blue R	Glycerol	L	814	RUBINOV and MOSTOVNIKOV (1967)
Methylene green	H_2SO_4	L	829	RUBINOV and MOSTOVNIKOV (1967)
Phthalocyanine	H_2SO_4	L	863	STEPANOV et al. (1967b)
Chloroaluminum phthalocyanine, aluminum phthalocyanine chloride	EtOH	L	755	SOROKIN and LANKARD (1966)
	DMSO	L	762	SOROKIN et al. (1967)
Magnesium phthalocyanine	Quinoline	L	759	STEPANOV et al. (1967b)
Chlorophyll	H_2SO_4	L	800	RUBINOV and MOSTOVNIKOV (1967)
3,3'-diethyloxacarbocyanine iodide, DOC iodide	Glycerol	F	541	MAEDA and MIYAZOE (1972)
3,3'-diethylthiacarbocyanine iodide, DTC iodide	Glycerol	F	625	MAEDA and MIYAZOE (1972)
1,3,3,1',3',3'-hexamethylindocarbo-cyanine iodide	Glycerol	F	614	MAEDA and MIYAZOE (1972)
Dicyanine	Glycerol	L	756	RUBINOV and MOSTOVNIKOV (1967)
	Quinoline	L	723, 752	RUBINOV and MOSTOVNIKOV (1967)
1,1'-diethyl-4,4'-carbocyanine bromide	Glycerol	L	754	MIYAZOE and MAEDA (1968)
1,1'-diethyl-4,4'-carbocyanine iodide, cryptocyanine	Glycerol	L	745	SPAETH and BORTFELD (1966)

Compound	Solvent		λ	Reference
1,1'-dimethyl-4,4'-carbocyanine iodide	Glycerol	L	749	Miyazoe and Maeda (1968)
3,3'-diethyloxadicarbocyanine iodide, DODC iodide	MeOH	L	658	Schäfer et al. (1967)
	DMSO	F	662	Maeda and Miyazoe (1972)
	MeOH, EG	C		Runge (1972)
3,3'-diethylthiadicarbocyanine iodide, DTDC iodide	MeOH	L	731	Schäfer et al. (1966)
	Acetone	L	711	Miyazoe and Maeda (1968)
	EG	L	728	Miyazoe and Maeda (1970)
	MeOH	C		Runge (1971)
3,3'-diethyl-5,5'-dimethoxy-6,6'-bis(methylmercapto)-10-methyl-thiadicarbocyanine bromide	DMSO	F	759	Maeda and Miyazoe (1972)
	EtOH	L	727–739	Derkacheva et al. (1968a)
3,3'-diethyl-10-chloro-(4,5,4',5'-dibenzo)-thiadicarbocyanine iodide	Acetone	L	774	Miyazoe and Maeda (1968)
3,3'-diethyl-10-chloro-(5,6,5',6'-dibenzo)-thiadicarbocyanine iodide	Acetone	L	714	Miyazoe and Maeda (1968)
1,3,3,1',3',3'-hexamethylindodicarbocyanine iodide	DMSO	F	740	Maeda and Miyazoe (1972)
1,1'-diethyl-2,2'-dicarbocyanine iodide	Glycerol	L	760–790	Spaeth and Bortfeld (1966)
	EG	L	812	Miyazoe and Maeda (1970)
	Glycerol	L	815	Miyazoe and Maeda (1968)
1,1'-diethyl-11-bromo-2,2'-dicarbocyanine iodide	Glycerol	L	745	Miyazoe and Maeda (1968)
1,1'-dimethyl-11-bromo-2,2'-dicarbocyanine iodide				
1,1'-diethyl-11-cyano-2,2'-dicarbocyanine fluoborate	MeOH	L	740	Schäfer et al. (1966)
	C_5H_5N	L	768	Bradley et al. (1968b)
1,1'-diethyl-11-acetoxy-2,2'-dicarbocyanine fluoborate	MeOH	L	797	Schäfer et al. (1966)
1,1'-diethyl-4,4'-dicarbocyanine iodide	EG	L	930	Miyazoe and Maeda (1970)
1,1'-diethyl-11-bromo-4,4'-dicarbocyanine iodide	MeOH	L	830	Miyazoe and Maeda (1968)
1,1'-diethyl-11-nitro-4,4'-dicarbocyanine fluoborate	MeOH	L	796	Schäfer et al. (1966)
	Acetone	L	814	Schäfer et al. (1966)
	DMF	L	815	Schäfer et al. (1966)
	C_5H_5N	L	821	Schäfer et al. (1966)

List of Laser Dyes (cont.)

Dye	Solvent[a]	Excit.[b]	Lasing Wavelength [nm]	Ref.
3,3'-diethyloxatricarbocyanine iodid, DOTC iodide	Acetone	L	742	Miyazoe and Maeda (1968)
	EtOH	L	718–739	Derkacheva et al. (1968a)
	EG	L	771	Miyazoe and Maeda (1970)
3,3'-dimethyloxatricarbocyanine iodide	Acetone	L	744	Miyazoe and Maeda (1968)
	DMSO	F	809	Maeda and Miyazoe (1972)
3,3'-diethylthiatricarbocyanine bromide, DTTC bromide	MeOH	L	813, 835	Schäfer et al. (1966)
	Acetone	L	808	Schäfer et al. (1966)
3,3'-diethylthiatricarbocyanine iodide, DTTC iodide	MeOH	L	816	Sorokin et al. (1966a)
	DMF	L	808	Sorokin et al. (1967)
	Glycerol	L	810	Sorokin et al. (1967)
	DMSO	L	816	Sorokin et al. (1967)
	Acetone	L	829	Miyazoe and Maeda (1968)
	DMSO	F	863, 889	Maeda and Miyazoe (1972)
3,3'-diethyl-11-methoxythiatricarbocyanine iodide	EtOH	L	773–798	Derkacheva et al. (1968a)
3,3'-diethyl-5,6,5',6'-tetramethoxythiatricarbocyanine iodide	Acetone	L	853	Miyazoe and Maeda (1968)
3,3'-diethyl-(4,5,4',5'-dibenzo)thiatricarbocyanine iodide	Acetone	L	860	Miyazoe and Maeda (1968)
3,3'-diethyl-6,7,6',7'-dibenzothiatricarbocyanine iodide	EtOH	L	824–853	Derkacheva et al. (1968a)
3,3'-diethyl-6,7,6',7'-dibenzo-11-methylthiatricarbocyanine iodide	EtOH	L	843–869	Derkacheva et al. (1968a)
3,3'-diethylselenatricarbocyanine iodide	Acetone	L	826	Miyazoe and Maeda (1968)
	EG	L	850	Miyazoe and Maeda (1970)
1,3,3,1',3',3'-hexamethylindotricarbocyanine iodide	EtOH	L	779–808	Derkacheva et al. (1968a)
	Acetone	L	819	Miyazoe and Maeda (1968)
		F	800	Furumoto and Ceccon (1970)
	EG	L	836	Miyazoe and Maeda (1970)
1,3,3,1',3',3'-hexamethyl-4,5,4',5'-dibenzoindotricarbocyanine perchlorate	EtOH	L	816–833	Derkacheva et al. (1968a)

1,1'-diethyl-2,2'-tricarbocyanine iodide	Acetone	L	898	MIYAZOE and MAEDA (1968)
	EtOH	L	886–898	DERKACHEVA et al. (1968a)
1,1'-diethyl-4,4'-tricarbocyanine iodide, xenocyanine	Acetone	L	1000	MIYAZOE and MAEDA (1968)
3,3'-diethyl-2,2'-(5,5'-dimethyl)-thiazolinotricarbocyanine iodide	Glycerol	L	717	MIYAZOE and MAEDA (1968)
3-ethyl-3'-methylthiathiazolinotri-carbocyanine iodide	EtOH	L	738–801	DERKACHEVA et al. (1968a)
3,3'-diethyl-12-ethylthiatetracarbo-cyanine iodide	EtOH	L	916–924	DERKACHEVA et al. (1968a)
1,1'-diethyl-13-acetoxy-2,2'-tetra-carbocyanine iodide	DMSO	L	1100	MIYAZOE and MAEDA (1970)
3,3'-diethyl-9,11:15,17-dineopenty-lenethiapentacarbocyanine iodide	$C_6H_5NO_2$	L	1093	DERKACHEVA et al. (1968b)
	MeOH	L	1100	SCHÄFER (1970)
	MeOH	L	1120	VARGA et al. (1968)
Pentacarbocyanine dye	$C_6H_5NO_2$	L	1175	VARGA et al. (1968)

[a] CH_2Cl_2 dichloromethane, CH_3CN acetonitrile, C_5H_5N pyridine, $C_5H_{11}OH$ isoamyl alcohol, $C_6H_4Cl_2$ o-dichlorobenzene, $C_6H_5CH_3$ toluene, $C_6H_5CF_3$ α,α,α-trifluorotoluene, $C_6H_5NO_2$ nitrobenzene, C_6H_6 benzene, $C_6H_{10}O$ cyclohexanone, C_6H_{12} cyclohexane, C_7H_{16} n-heptane, DCE dichloroethane, DMF N,N-dimethylformamide, DMSO methyl sulfoxide, DPA N,N-dipropylacetamide, EG ethylene glycol, EtOH ethanol, HFIP hexafluoroisopropanol, H_2O water, H_2SO_4 sulfuric acid, IP isopentane, MCH methylcyclohexane, MeCs methyl cellosolve, MeOD mono-deuterated methanol, MeOH methanol, NMP 1-methyl-2-pyrrolidinone, PMMA polymethylmethacrylate, PSt polystyrene, TFE trifluoroethanol, THF tetrahydrofuran.

[b] L laser-pumped, F flashlamp-pumped, C continuous (laser-pumped).

5. Applications of Dye Lasers

Theodor W. Hänsch

With 26 Figures

The development of various types of tunable dye lasers in the past few years has led to remarkable progress, and it does not seem premature to predict that dye lasers will have a considerable impact on a variety of areas of science and technology. To illustrate the potential and versatility of organic dye lasers, this last chapter will review a number of recently reported or suggested dye-laser applications. Naturally, most of these applications have so far been confined to the research laboratory. Technical applications of dye lasers have not yet had time to mature beyond the stage of feasibility studies. Many new and perhaps surprising applications of dye lasers will certainly be found in the future.

Properties of Dye Lasers

Although organic dye lasers have been discussed in detail in the preceeding chapters, it appears worthwhile to summarize some of their characteristic properties which are useful for applications.

Considering that they are in competition with various somewhat more advanced fixed-frequency lasers, it is not surprising that most applications of dye lasers reported so far depend on their unique capability to tune the laser wavelength continuously over a wide range, and to channel the laser energy without substantial loss into a narrow spectral line. Pulsed dye-laser radiation can now be generated at any wavelength from the near-ultraviolet (340 nm) to the near-infrared (1200 nm) using both flashlamp and laser pumping. Continuous dye-laser emission has been obtained throughout the visible spectrum, at least in the laboratory, using excitation with visible and ultraviolet lines of an argon-ion laser. The accessible wavelength range can be extended in both directions by frequency mixing in nonlinear optical materials. Dye lasers have not been breaking any records as far as their spectral linewidth or temporal coherence is concerned, because wideband tunability and extremely narrow bandwidth are somewhat contradictory requirements, but the progress which has been achieved very recently is nonetheless impressive. Bandwidths between 1 and 10 pm can be obtained quite easily from pulsed or continuous dye lasers with the help of a wavelength-selective cavity element, such as a grating, prism, or interference filter. Narrower lines can be produced with multiple dispersive elements. A linewidth of less than 1 pm and laser oscillation in a single-cavity mode can be enforced in flashlamp-pumped dye lasers with multiple tilted Fabry-Perot etalons (BRADLEY et al., 1971) or with Lyot filters and a diffraction grating (WALTHER and HALL, 1970). A bandwith of 0.4 pm or 300 MHz has been obtained from a repetitively pulsed dye laser, side-pumped by a nitrogen laser, by using a diffraction grating, an intracavity beam-expanding telescope, and a Fabry-Perot etalon (HÄNSCH, 1972). This linewidth is not too far above the Fourier transform limit imposed by the short nanosecond pulse width.

Considerably narrower lines can in principle be generated with continuous dye lasers. Single-mode operation and bandwidths of the order of 10—50 MHz have been achieved with intracavity prisms and Fabry-Perot etalons (HERCHER and PIKE; SCHUDA et al., 1973; HARTIG and WALTHER, 1973). BARGER et al. [1973] have been able to reduce the linewidth even further to a few MHz by actively controlling the laser cavity length with a rapid electronic feedback system and using a passive optical resonator as frequency standard. Residual fluctuations of the line center relative to this standard are less than 50 kHz for short times (20 μsec) and less than 100 Hz for long times (10 sec). Extremely narrowband dye lasers with multiple wavelength selectors have so far only limited tuning ranges, but considerable engineering efforts are being invested in the development of dye lasers which can be tuned continuously and smoothly over wide spectral regions. Several workers have successfully demonstrated schemes for stabilizing the dye-laser frequency to atomic or molecular resonance lines (SOROKIN et al., 1969; BÖLGER and WEYSENFELD, 1972; KLEIN, 1972; WALTHER and HARTIG, 1973).

It is well known that lasers can be greatly superior to conventional light sources in their spatial coherence which determines, for instance, how well the radiation can be collimated, or how tightly it can be focused. The spatial coherence of dye lasers can be comparable to that of gas lasers, i.e. very close to the theoretical limit imposed by the wave nature of light. The oscillation of laser-pumped pulsed or continuous dye lasers can be restricted to the fundamental transverse TEM_{00} cavity mode. Even side pumping can provide a near diffraction-limited beam if the laser cavity is properly designed (HÄNSCH, 1972). Difficulties are sometimes encountered with flashlamp-pumped dye lasers because irregular filaments in the flashtubes can result in poorly defined pump geometry.

Other important parameters for dye-laser applications are the available light energy and power. The energies of pulsed dye lasers range from a few microjoules to more than 10 joules per pulse, and the peak powers range from milliwatts to more than 100 megawatts. While flashlamp-pumped dye lasers with much higher output energies can certainly be built, it appears unlikely that they can provide the very high intensities of Q-switched solid state lasers, because the short lifetime of the excited singlet state prevents the storage of pump energy in the dye medium for periods longer than a few nanoseconds. If tunable radiation at extremely high power is required, it is possible to use dye lasers, pumped by Q-switched solid state lasers, as efficient wavelength converters. Conversion efficiencies of up to 50% can be obtained with certain dyes under pulsed excitation. Continuous dye lasers, pumped by commercial argon-ion lasers, can provide powers from a few milliwatts to several watts with an efficiency exceeding 10%. The efficiency is generally lower if very narrow linewidths are required.

Some laser applications require short pulse durations. Dye lasers are extremely versatile in this respect. Flashlamp-pumped devices can provide pulses ranging from a fraction of a microsecond to several milliseconds. Laser-pumped dye lasers often operate in the region from 1—200 nsec. Pulse widths down to a few picoseconds can be obtained by mode-locking, as discussed in Chapter 3.

Dye lasers which do not have to meet extreme requirements can be technologically simple. The active medium is generally inexpensive and easily available, and problems of heat dissipation or material damage are easily avoided by circulating the liquid laser medium. However, many of these potential advantages are presently defeated by the lack of an efficient, reliable and inexpensive pump source. Flashlamps permit simple designs of dye laser, but their short lifetime rules out economical long-term operation at high repetition rates. Pump lasers such as the pulsed nitrogen laser can be very reliable, but they are rather inefficient and often expensive. Better pump sources, such as suitable light-emitting semiconductor diodes, would clearly be very desirable. If such sources were developed, dye lasers could compete successfully with solid-state lasers, gas lasers or semiconductor lasers in many applications which do not require wavelength tunability.

Laser Applications

Innumerable applications of laser light have been proposed, developed and demonstrated, since SCHAWLOW and TOWNES published their first paper on optical masers in 1958. Certain applications have reached a stage of technical maturity even in this relatively short time. The unique properties of laser light have opened the door to several new fields which are being explored with vigorous activity, including nonlinear optics, the physics of ultrashort pulses, and holography. Such venerable classical fields as optics, spectroscopy and metrology have experienced a forceful rejuvenescence. The new topics of coherent optics and quantum optics have attracted many researchers. Very important breakthroughs have been achieved in spectroscopy. Studies of light scattering, in particular Raman spectroscopy, have been virtually revolutionized by the advent of laser sources. Many interesting and often powerful new nonlinear spectroscopic phenomena have been demonstrated despite the limitations of fixed-frequency lasers, in particular stimulated light-scattering processes, several multiphoton processes, selective saturation of resonant transitions, and various coherent transient effects. Advances in nonlinear laser spectroscopy are creating new standards of length and possibly time, too. The upper limit for frequency measurements has been advanced from the microwave region into the near-infrared. Lasers are already quite widely used for alignment and ranging. Accurate interferometric measurements of length have become possible over long distances. Lasers are also used for velocity measurements, using the Doppler effect. The potential of lasers for communications and data transmission was early recognized but is still far from being fully utilized. The processing of various materials with laser light, in particular drilling, cutting or welding, finds increasing acceptance in industrial production. The successful uses of lasers for eye surgery are widely publicized. Large efforts are under way in different countries to attempt the initiation of nuclear fusion with powerful laser pulses.

It is impossible within our limited space to discuss or even to list all the laser applications which have been reported in literally thousands of publications. The reader is referred to several rather comprehensive collections of laser abstracts [e.g., TOMIYASU (1968) or Journal of Current Laser Abstracts (since 1964)]. A bibliography of laser books, including many monographs on laser applications, has been compiled by SIEGMAN (1971). Very recent publications include books on laser applications in industry (SMITH, 1971; BEESLY, 1971; ROSS, 1971; CHARSCHAN, 1972) and in medicine and biology (WOLBARSHT, 1971; GOLDMAN and ROCKWELL, 1971). There is little doubt that dye lasers can perform many of the reported laser applications; we shall, however, restrict the discussions in this chapter to those applications which have already been demonstrated or suggested for dye lasers.

5.1. Light Absorption and Scattering

Dye lasers offer several obvious advantages over conventional light sources and monochromators for the study of optical absorption spectra. The high intensity, good directionality, and narrow linewidth of laser light have also led to numerous important advances in the scattering of light from matter, both for practical diagnostic applications and for basic research. Tunable dye lasers are naturally of particular value whenever the scattering phenomenon depends critically on the light wavelength. Several interesting applications of dye lasers in Raman scattering have already been reported, but perhaps the most promising use is the excitation of resonant scattering from atoms and molecules, with applications ranging from trace analysis in the laboratory to the remote detection of atmospheric constituents.

One noteworthy attraction of dye lasers for such applications is the possibility of electronic wavelength tuning, using for example acousto-optic filters (TAYLOR et al., 1971), Lyot filters (WALTHER and HALL, 1970) or piezoelectrically tunable Fabry-Perot interferometers inside the dye-laser cavity. Such devices enable the laser wavelength to be scanned automatically or very rapidly or electronically locked the wavelength to a particular absorption line.

The nanosecond pulses of laser-pumped dye lasers or the picosecond pulses of mode-locked dye lasers permit time-resolved measurements which can be extremely valuable for the investigation of transient species or for the study of relaxation phenomena and chemical reaction kinetics. The only serious limitation of dye lasers appears to be their somewhat limited tuning range.

5.1.1. Absorption Spectroscopy

The study of optical absorption spectra of atomic or molecular gases, liquids, and solids is a widely used analytic technique. Not only can dye lasers replace conventional lamps and monochromators for the mapping of absorption spectra, but their narrow linewidth permits superior spectral resolution and their spatial coherence facilitates the study of very small or very long samples and the use of multiple light passes of well defined geometry. The high spectral brightness of dye-laser radiation can provide a good signal-to-noise ratio even in the limit of very high resolution and short measuring times.

Atomic absorption spectroscopy is an analytic technique of particular importance, and comprehensive review articles have been written about this field (WINEFORDNER and VICKERS, 1970). Commercial absorption spectrometers require a separate hollow-cathode spectral lamp for each metallic element. Tunable dye lasers offer attractive possibilities for simplification and automation, provided their tuning range can be extended further into the ultraviolet, down to about 200 nm.

Certain nonreproducible or short-living samples require the simultaneous recording of the entire absorption spectrum, using a bright continuum light source, a spectrograph and a photographic emulsion or the storage photocathode of a TV image tube. High spectral resolution and short observation times call for extremely bright sources. Ideal for such applications are laser-pumped pulsed dye lasers without optical cavity, operating in the "superradiant" mode, i.e. emitting emplified spontaneous fluorescence in near axial direction. They can easily reach intensities corresponding to blackbody radiation temperatures of $10^{10}\,°K$ and they provide a smooth, structureless spectrum. Such dye lasers have, for instance, been used successfully to measure the absorption of optically excited rubidium atoms (BRADLEY et al., 1970).

5.1.2. Selective Dye-Laser Quenching

Broadband dye lasers with optical cavity have been used by several workers in a novel, unconventional way as nonlinear active sensors for small selective optical extinctions. Such dye lasers typically operate in a 2—20 nm wide band near the gain maximum of the dye solution. If a discrete absorber is placed inside the laser cavity, it reduces or quenches the laser emission at the absorbed wavelength and channels the dye-laser energy into the intermediate spectral regions.

An increase in sensitivity of two to three orders of magnitude as compared to a conventional single-pass measurement was observed for absorbing diatomic iodine vapor or dilute $Eu(NO_3)_3$ solutions inside the cavity of a flashlamp-pumped dye laser (PETERSON et al., 1971; KELLER et al., 1972). The same technique made it possible to detect the weak absorption of Sr atoms and Ba ions provided by a laminar flame burner inside the laser resonator, as demonstrated in Fig. 5.1 (THRASH et al., 1971).

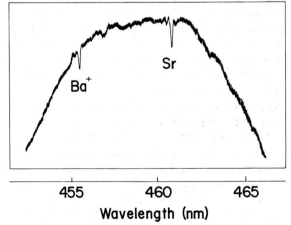

Fig. 5.1. Spectral output of a broadband flashlamp-pumped coumarin dye laser with an acetylene-air flame doped with Ba and Sr inside the laser cavity. (THRASH et al., 1971)

An increase in sensitivity of 10^5 was measured by placing an absorption cell containing I_2 vapor inside the cavity of a cw dye laser and observing the selective laser quenching by monitoring the fluorescence of a second external iodine cell (Hänsch et al., 1972). For complete quenching it was sufficient to allow some iodine vapor in air to diffuse into the optical cavity by placing an iodine crystal near the loss region. Figure 5.2 illustrates a rather amusing demonstration. It shows two

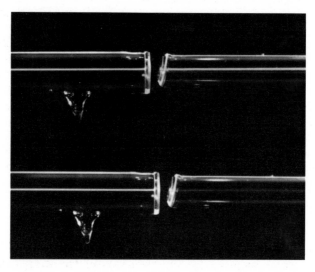

Fig. 5.2. Two fluorescence cells with I_2^{127} (right) and I_2^{129} (left) vapor outside the cacity of a broadband cw dye laser (Top: without intracavity absorption; bottom: with a trace of I_2^{127} vapor inside the laser cavity) (Hänsch et al., 1972)

fluorescence cells, each filled with a different iodine isotope, outside the dye-laser cavity. Without intracavity absorption, both cells show an equally bright fluorescent sidelight. A trace of I_2^{127} inside the resonator completely quenches the fluorescence of the external cell containing the same isotope, while the fluorescence of the other cell appears almost unchanged because its absorption lines are slightly displaced by the isotope shift. This method could provide a convenient means for rapid quantitative chemical or isotope analysis with ultrahigh sensitivity and without the need for any spectrophotometric device.

The opposite effect can be observed if a negative absorber, i.e. a medium with small-wavelength selective gain, is placed inside the dye-laser cavity. This was demonstrated by Klein (1972) with a cw dye laser, internally pumped by an argon-ion laser: the dye laser tended to oscillate in narrow lines, corresponding to inverted atomic transitions in the pump discharge which are normally too weak to reach laser threshold. The effect can be utilized to lock the dye-laser frequency to an inverted atomic transition.

The large sensitivity of a broadband dye laser to selective intracavity absorption can be explained by means of a rate equation model (HÄNSCH et al., 1972) taking into account the spectrally homogeneous but spatially inhomogeneous saturation of the dye-laser gain by the different cavity modes. Under steady-state conditions, the intensity of each mode adjusts so that the saturated single-pass gain equals the losses. It is obvious that a large absorption enhancement factor can be obtained, even with a single mode laser, if it operates sufficiently close to threshold. An additional increase in sensitivity by a factor equal to twice the number of oscillating cavity modes is predicted for a broadband dye laser, because all oscillating modes participate in the saturation of the gain, and a large drop in intensity for an individual mode is required for a small increase in laser gain. A different model has been suggested by KELLER et al. (1972), neglecting the spatial inhomogeneity of laser saturation and ascribing the coexistence of different modes in a pulsed dye laser to transient effects.

5.1.3. Raman Scattering

When molecules are irradiated with monochromatic light, some of the scattered light is shifted from the frequency of the incident light by resonance frequencies of the scatterer. This inelastic light scattering is known as Raman scattering and has become a widely used tool for the study of molecular vibration frequencies. It is particularly valuable because, due to its different selection rules, it complements infrared absorption spectroscopy. The advent of various fixed-frequency lasers has revolutionized the instrumentation of this analytic technique and has stimulated intensive research efforts, in particular the study of Raman scattering from lattice vibrations, spin waves or electronic excitation in solids. An account of laser Raman spectroscopy has been given in several monographs (SUSCHINSKII, 1972; GILSON, 1970; TOBIN, 1971).

The intensities of Raman-scattered light are usually very weak (typical scattering cross sections are of the order of 10^{-28} cm^2 or less) and double or triple monochromators with a high rejection ratio for unwanted frequencies are required to suppress the background of unshifted scattered laser light. A tunable dye laser would cut out all the complications of precision tracking of these monochromators. In principle, all that is required is a fixed-frequency filter in front of the photodetector, then the Raman spectrum could be recorded by changing the frequency of the incident laser light. The feasibility of this arrangement has been demonstrated by McNice (1972), using a stack of narrowband interference filters and a flashlamp-pumped dye laser. An additional advantage of this scheme is that spectral filters of high transmittance, such as Fabry-Perot interferometers, can be used without any need to guide the scattered light into the narrow slit of a spectrometer. Extremely high resolution and a wide acceptance angle could be realized by using the Raman scattered light to excite atomic line fluorescence in a suitably designed scattering cell near the detector.

A rather efficient way to suppress elastically scattered laser light is by use in front of the spectrometer of a resonant absorber whose frequency coincides with the laser frequency. A molecular iodine filter can provide a rejection ratio of better than 10^7 for unshifted argon-ion laser light (514.5 nm) (Devlin et al., 1971). Tunable dye lasers permit the use of absorbing atomic vapors with a much simpler line spectrum (e.g., Na) which would avoid any interference with the Raman spectrum.

It has long been known that a marked increase in the Raman-scattering cross section is observed, when the incident light frequency or the Raman-shifted frequency approaches the resonance of one or more electronic transitions of the scatterer. At an exact resonance, this "resonance Raman scattering" turns into resonance fluorescence, and there is no sharp distinction between the two phenomena. A resonant enhancement in the Raman-scattering cross section of molecules is also observed when the excitation is near or above the dissociation limit and a continuum of states must be integrated. Considerable theoretical and enperimental interest has recently been focused on resonance Raman scattering. Tunable dye lasers are already proving ideal light sources for such studies. Damen and Shaw (1971) have used a flashlamp-pumped dye laser to measure the enhancement in scattering cross section in CdS due to bound exciton intermediate states. Yu and Shen (1972) have measured resonance Raman scattering in InSb near the E_1 transition, using a continuous dye laser.

Closely related to studies of resonant Raman scattering is the resonant excitation of luminescence in solids. Petroff et al. (1972) have demonstrated the superiority of selective excitation with a cw dye laser in an investigation of the luminescence of CuO_2 crystals at low temperatures. They were able to distinguish clearly between different excitons and to observe new resonance peaks. They could also rule out the existence of excitonic molecules, which had been suggested previously.

The increasing use of dye lasers for resonance Raman scattering promises many interesting results in this field.

5.1.4. Resonance Fluorescence

The wavelength tunability of dye lasers is of particular value for the excitation of resonance fluorescence from atoms and molecules. Typical Doppler-broadened resonance lines have a width of the order of 1 pm, i.e. above the limits of pulsed or continuous dye lasers, so that single resonance lines may readily be selectively excited.

Sodium, with its two strong yellow resonance lines near the gain maximum of rhodamine 6G, has been the favorite substance in numerous experiments undertaken to study the feasibility of atomic fluorescence

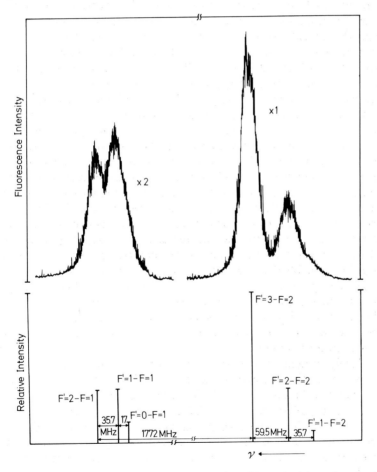

Fig. 5.3. High-resolution spectrum of the Na D_2 line with resolved hyperfine structure, measured in fluorescence with a collimated atomic beam excited by a cw dye laser. (HARTIG and WALTHER, 1973)

excitation with tunable dye lasers (e.g. WALTHER and HALL, 1970; YAMAGUCHI et al., 1969). Other studies of dye-laser-excited atomic fluorescence include the observation of atomic flame fluorescence of the green barium resonance line by DENTON and MALMSTADT (1971) and the observation of resonance scattering from excited helium atoms in a cold plasma by BURRELL and KUNZE (1972). The latter experiment was used to study collisional and radiative transfer rates.

The excitation of resonance fluorescence from atomic and molecular beams opens up numerous interesting applications. Such beams are widely used in physics research, from the measurement of radiofrequency transitions between hyperfine levels to the study of collision processes and chemical reactions. One major problem is the sensitive and selective detection of neutral particles. Dye lasers will certainly find increasing use for this purpose.

The scattering of dye-laser radiation from an atomic beam moving at right angles to it can provide a means for very-high-resolution optical spectroscopy without blurring of the line shape and structure by Doppler broadening due to the thermal motion of the atoms. HARTIG and WALTHER (1973) have observed linewidths as narrow as 30 MHz for the hyperfine components of the NaD lines by using an atomic beam with a collimation of 1:550 and a cw dye laser of less than 20 MHz bandwidth and 10 mW power. Single-mode operation was achieved with the help of an intracavity etalon and continuous-frequency tuning over 600 MHz by piezoelectric variation of the cavity length. An electro-optically variable transmission filter controlled the intensity within 10^{-3}. Figure 5.3 shows a spectrum of the NaD_2 line with resolved hyperfine structure, as obtained with this apparatus. The beam fluorescence could also be utilized to lock the dye-laser frequency to one of the Na hyperfine components. A similar direct optical observation of the Na hyperfine structure using a cw dye laser and an atomic beam was reported by SCHUDA et al. (1973).

5.1.5. Trace Analysis, Pollution Detection

Trace analysis is one valuable application of fluorescence spectroscopy. Atomic fluorescence spectroscopy with conventional hollow-cathode lamps is potentially one of the most sensitive techniques of analytic spectroscopy (WINEFORDNER et al., 1971, 1972). The detection limits are determined by the low scattering intensities and background scattering, and are typically on the order of 10^{-16} to 10^{-10} grams or 10^7 to 10^{13} atoms. Dye lasers are much superior to hollow-cathode lamps in their spectral brightness and solid angle of emission and promise a much improved performance. Moreover, their tunability makes this technique applicable to rare or excited atoms, ions and molecules for which strong lamps have not so far been available.

KUHL and MAROWSKY (1971) have reported the detection of Na atoms in concentrations as low as 0.003 ng/cm^3, using excitation with a flash-lamp-pumped dye laser. Much higher sensitivities should be possible with cw dye lasers. In principle, it should even be possible to detect the resonance radiation scattered by a single atom. One atom is capable of repeated absorption and reemission of photons, thus it should permit repeated "nondestructive" observation over an extended period of time, provided interfering processes such as two-photon ionization are negligible. A two-level atom irradiated by intense resonant laser light can spend at most one half of its time in the excited state, i.e. the scattered intensity per atom is at most $A\,h\nu/2$, where A is the inverse radiative lifetime. For the Na D$_2$ resonance line with an upper-state lifetime of 16 nsec one expects $2 \cdot 10^7$ scattered photons per second from one atom, if the intensity of the incident resonant laser radiation equals the saturation intensity $I_{sat} = 21$ mW/cm^2. The moderate power requirement is due to the large resonance-scattering cross section, whose diameter can approach the wavelength of light.

The possibility of detecting atmospheric pollutants with the help of tunable lasers has recently attracted attention (MELNGAILIS, 1972; HINKLEY and KELLEY, 1971). Atomic and molecular fluorescence is unfortunately often quenched by the presence of foreign gases at high pressure. Resonance Raman scattering is much less susceptible to collision quenching because it involves "virtual" intermediate excited states of much shorter lifetime. The measurement by means of tunable lasers of absorption in atmospheric trace constituents can also reach considerable levels of sensitivity because it is possible to utilize very long optical paths. Most pollutants of practical interest have absorption bands in the infrared at wavelengths between 2 and 10 μm, which are not directly accessible to organic dye lasers. Dye lasers may nonetheless become important tools for pollution detection and monitoring, because it is possible to generate tunable infrared radiation by mixing the frequencies of two dye lasers in a nonlinear optical crystal (see Section 5.5.2).

5.1.6. LIDAR, Resonant Atmospheric Scattering

Since pulsed dye lasers can provide high intensities, they can be used in a radarlike fashion for the remote observation of resonance fluorescence from atmospheric trace constituents at high altitudes. An apparatus

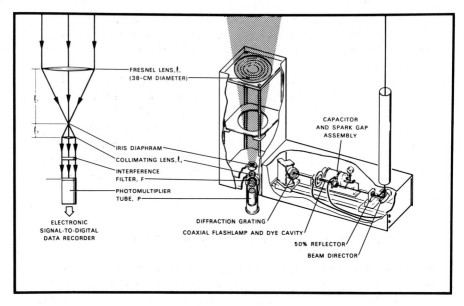

Fig. 5.4. Scheme of tunable-laser radar apparatus for resonant backscattering from the upper atmosphere. (GRAMS and WYMAN, 1972)

suitable for such LIDAR (Light Detection and Ranging) measurements is shown in Fig. 5.4. The light of a flashlamp-pumped tunable dye laser is emitted through a collimating telescope, and a second telescope of large aperture is used to guide the backscattered light into the receiving photomultiplier, BOWMAN et al. (1968), and SANDFORD and GIBSON (1970) have successfully used a flashlamp-pumped dye laser of 2 mJ pulse energy, interferometrically tuned to one of the Na D resonance lines (GIBSON, 1969) to detect and map patches of free Na vapor in the earth's atmosphere at an altitude of approximately 90 km. Systematic twilight and night-time measurements (SANDFORD and GIBSON, 1970) revealed abundances of several 10^9 Na atoms/m^3 in patches of 2—3 km thickness and 10—200 km horizontal dimensions. SCHULER et al. (1971) operating with 200 mJ dye laser pulses reported somewhat lower Na concentrations. An extension of such measurements, which is presently being pursued by several groups, may provide information on the origin of the atmospheric sodium and on chemical and transport mechanisms in the upper atmosphere. Other atoms, ions or molecules should also be detectable via resonant backscattering (GIBSON, 1969), but most of them will require considerably stronger dye-laser pulses. The detection of free OH radicals is of particular interest, because this radical could be controlling the worldwide conversion of CO into CO_2. Laboratory experiments by BAARDSEN and TERHUNE (1972) have already confirmed that it is possible to detect OH in concentrations lower than 10^{12} molecules/cm^3 by excitation with a frequency-doubled dye laser, operating at 282.2 nm.

Other workers have demonstrated the feasibility of using LIDAR measurements for the remote observation of Raman scattering from atmospheric constituents or pollutants (MELFI, 1972). Such measurements have so far been confined to a relatively short ranges, e.g. monitoring of industrial smoke stacks, and most have been performed with fixed-frequency solid-state lasers. It is obvious that measurements of resonant Raman scattering with tunable dye lasers can result in considerably increased sensitivities.

5.1.7. Measurements of Lifetimes and Relaxation Rates

Pulsed dye lasers are ideally suited for the measurement of lifetimes of electronically excited states in atoms or molecules. The narrow bandwidth permits the selective excitation of the energy levels of interest, and the lifetime can be obtained directly from measurements of the exponential decay of the subsequent spontaneous emission or resonance fluorescence. Several authors have reported measuring lifetimes of simple molecules, using dye lasers with a relatively broad bandwidth of about a nanometer that do not distinguish between molecular rotational states. SACKETT and YARDLEY (1970) have measured lifetimes of 70—82 μsec in NO_2 with a flashlamp-pumped dye laser, operating between 440 and 490 nm. SAKURAI and CAPELLE (1970) performed similar measurements, using a dye laser pumped by a pulsed nitrogen laser, and obtained somewhat shorter lifetimes. The 10 nsec pulse length of this dye laser is able to determine lifetimes of less than 0.1 μsec. Such measurements have been reported for BaO (JOHNSON and BROIDA, 1970), for I_2 (SAKURAI et al., 1971) and for Br_2 (CAPELLE et al., 1971). ERDMANN et al. (1972) have described a versatile apparatus for lifetime measurements, featuring a flashlamp-pumped dye laser of 1 pm bandwidth whose oscillation can be interrupted within a few nanoseconds by means of an electronically tunable Lyot filter inside the laser cavity. Such a scheme can reduce distorting oscillations of the intensity decay curves by quantum beats due to coherently excited, closely spaced levels (see Section 5.4.2). Figure 5.5 shows the results of a test measurement, performed with the $3^2 P_{1/2}$ state of atomic sodium. The fluorescence intensity at different time intervals after quenching the laser oscillation has been recorded with gated integrators. The logarithmic display confirms the exponential decay and yields a rather satisfactory lifetime value of

16 nsec. BEYER and LINEBERGER (1973) have used a nitrogen-laser-pumped dye laser, together with a computerized data acquisition system, to excite the 0—0 band of *cis*-glyoxal near 487.5 nm, at pressures from 30 to 150 m Torr and to measure lifetimes and quenching rate constants of the first excited singlet states of *cis* and *trans*-glyoxal.

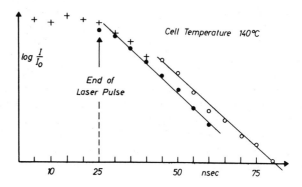

Fig. 5.5. Measurement of atomic lifetimes with a pulsed dye laser. Logarithm of the fluorescence intensity versus time lag after laser switch-off. (ERDMANN et al., 1972)

Mode-locked picosecond dye lasers permit an extension of lifetime and relaxation measurements to very short times, such as the subnanosecond rotational relaxation times of organic molecules in a liquid environment. The feasibility of such measurements has already been demonstrated, using fixed-frequency mode-locked solid-state lasers (EISENTHAL and DREXHAGE, 1969).

5.1.8. Resonant Radiation Pressure, Atomic Beam Deflection

Because of the large resonant light-scattering cross section of atoms and molecules, a substantial radiation pressure force can be exerted on such neutral particles, using a resonant dye-laser radiation field. ASHKIN (1970) has pointed out that this force could, for instance, be used to selectively pump gas atoms against significant gas pressures, or to trap and confine gas atoms. ASHKIN also proposed to use a laser radiation field of cylindrical geometry to deflect atomic beams in circular orbits. The direction of light propagation would always be perpendicular to the atomic velocity, so that the atoms are not Doppler-shifted out of resonance. Such a scheme could be utilized to separate atoms of specific isotopic species or hyperfine level, or it could serve as a velocity analyzer for neutral particles. This latter application would benefit from the fact that the radiation pressure force reaches an intensity-independent value in the limit of very high laser intensity or complete saturation of the resonant atomic transition. The radiation pressure force then can constitute a central force field with focusing properties analog to the cylindrical electric field in electron-velocity analyzers.

An experimental observation of the deflection of a beam of Na atoms by exchange of linear momentum with resonant photons was reported as early as 1933, by FRISCH who used a Na discharge lamp. The effect was very small, however, since only one third of the atoms in the beam were excited by the light. SCHIEDER et al. (1972) have used a cw dye laser of 10 mW power and less than 50 MHz bandwidth to deflect a sharply collimated sodium beam. The laser was tuned to the hyperfine transition $^2S_{1/2}, F = 2 - ^2P_{3/2}, F = 3$ of the D_2 resonance line, so that the radiative selection rules would prevent the depletion of the ground state by optical pumping between hyperfine levels. About 30% of the sodium atoms were deflected in these preliminary experiments, with a linear momentum transfer corresponding to 60 excitations per atom.

5.2. Selective Excitations

The intense radiation of tunable dye lasers can be applied to excite selected energy states of atoms, ions, molecules or solids. It is often possible to excite a substantial fraction of the resonant absorbers. An infinite temperature equilibrium can be momentarily established between the upper and lower levels in the limit of complete saturation. This possibility is of substantial interest for a number of practical applications, ranging from plasma diagnostics (MEASURES, 1968; DEWEY, 1971) to the optical excitation of new laser transitions.

5.2.1. Optical Pumping

It is well known that atomic vapors can be polarized by optical pumping, i.e. by the repeated absorption and reemission of polarized resonance radiation (HAPPER, 1972). The selection rules of the absorption and emission cycle lead to the preferential population of particular orientational (hyperfine) states of the stable ground level. Optical pumping is widely used for precision spectroscopy, for optical magnetometers, for atomic clocks, and for the preparation of polarized gas samples, e.g. as targets in nuclear physics. The construction of a sufficiently strong spectral lamp and of a suitable spectral line filter often constitutes a major experimental problem. Tunable dye lasers

appear to offer an ideal solution to these problems, and they will
certainly permit an extension of optical pumping techniques to rare
atoms, ions, and molecules for which strong spectral lamps are not
available. Dye lasers should also permit the optical polarization of very
dense vapors which are opaque at exact resonance but sufficiently
transparent at the wing of a resonance line. A reasonable pumping effect
might be achieved already with pulsed dye lasers, despite the limited
number of pumping cycles during the laser pulse length. But the ideal
pump source would certainly be a widely tunable cw dye laser.

RUFF et al. (1971) have reported the polarization of Na vapor by
optical pumping with a cw dye laser. The polarization time constant is
at least three orders of magnitude smaller than that obtained with
conventional lamps.

5.2.2. Excited-State Spectroscopy

Several workers have demonstrated that dye lasers can provide the
excitation required for absorption spectroscopy from excited atomic
states. MCILRATH (1969) has used a pulsed dye laser, pumped by a
frequency-doubled Nd: glass laser, to pump the intercombination transition in atomic Ca vapor at 657.3 nm. The laser pulses of 30 mJ energy
and 1 cm^{-1} bandwidth were sufficient to equalize the population of the

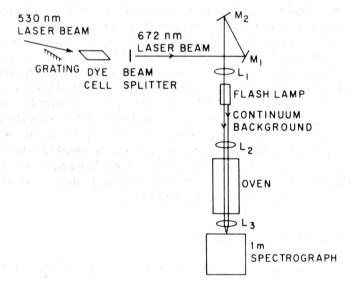

Fig. 5.6. Experimental setup for absorption spectroscopy from excited states in dye-laser-
pumped calcium. (MCILRATH, 1969)

ground state and the excited metastable $3P(4s4p)$ level. Absorption spectra from the excited level were subsequently recorded by means of a flashlamp continuum. A scheme of the apparatus is shown in Fig. 5.6.

BRADLEY et al. (1970) applied a ruby-pumped dye laser with an output of 3 mJ per pulse at 775.8 nm and 1 cm^{-1} bandwidth to selectively populate the $5P_{3/2}$ level of atomic rubidium. A second broadband dye laser provided the continuum for an absorption measurement at the excited-state absorption line $5P_{3/2} - 5D_{5/2}$ at 775.8 nm.

A similar scheme was later utilized to measure photoionization cross sections for excited magnesium atoms which are of particular interest for the interpretation of astronomical spectra (BRADLEY et al., 1972). Magnesium atoms in an oven were selectively excited to the $3s$, $3p$, ^1P state by a megawatt second harmonic beam of a dye laser, tuned to 285.55 nm and with a bandwidth of 0.04 cm^{-1}. The dye laser was side-pumped by a frequency-doubled Nd : glass laser. The UV continuum emitted by a scintillator dye or a second, probing dye-laser beam of different frequency were used to measure broadband and narrowband absorption from the excited magnesium, corresponding to different states of the magnesium ion.

5.2.3. Photomagnetism

Intensive selective excitation with tunable dye-laser light is revealing novel possibilities in solid-state spectroscopy. An interesting example is a study of photomagnetism, reported by HOLZRICHTER et al. (1971). These authors have used a flashlamp-pumped dye laser, tuned to 542.0 nm, and with an output of about 2 mJ at a bandwidth of 0.1 cm^{-1}, to excite both excitons and magnons in a crystal of antiferromagnetic manganese fluoride near the liquid helium temperature. The magnetization corresponding to the spin change of the manganese ions is sufficiently strong to be observed with a pick-up coil around the crystal. Stressing the crystal enables the two exciton lines, corresponding to the two anti-parallel sublattices, to be split, and the laser can be tuned to excite the spin in one or the other sublattice. With some detuning it was also possible to selectively excite spin-wave sidebands and to measure relaxation times for both excitons and magnons.

5.2.4. Pumping of Laser Transitions

Strong selective excitations with tunable lasers is an excellent pumping mechanism to obtain population inversion and laser action on atomic or molecular transitions. Such optically pumped lasers not only provide interesting objects for nonlinear-spectroscopic investigation, but they also can serve as precise frequency markers or secondary wavelength standards, and they may even be useful as coherent light sources at new wavelengths.

SOROKIN and LANKARD (1969) have used a pulsed dye laser, pumped transversely by the second harmonic beam of a Q-switched ruby laser, with 300 kW peak power and 30 cm^{-1} bandwidth, to pump the singlet resonance line at 460.7 nm in Sr vapor in a helium buffer gas. They obtained infrared laser action from various triplet transitions in Sr. In this case, following the optical excitation, the upper states are populated by inelastic collision processes.

Other workers have reported laser action in atomic vapors after selective resonant two-photon excitation with dye-laser light. KOROLEV et al. (1970) have observed directed stimulated emission in Rb vapor at the transition $5^2S_{1/2} - 6^2P_{3/2}$ (420.2 nm) and $5_2S_{1/2} - 6^2P_{1/2}$ (421.5 nm) when the vapor was pumped with a ruby-pumped dye laser operating in the range 773.0—790.0 nm. The pump threshold drops to a sharp minimum of about 100 kW/cm^2 if the dye laser is tuned to the two-photon resonance $5^2S_{1/2} - 5^2D_{3/2, 5/2}$ near 777.8 nm. Spontaneous transitions from the 5D levels seem to be responsible for the population inversion on the laser transitions. AGOSTINI et al. (1972) have reported similar results for K vapor.

SOROKIN and LANKARD (1971) have obtained infrared laser action from K, Rb, and Cs vapor in a 2—4 m long heat-pipe oven after optical excitation with various strong pulsed lasers, including laser-pumped dye lasers. The mechanism of excitation in these experiments was two-photon absorption by diatomic molecules, followed by photodissociation into the excited atomic states.

Numerous visible and near infrared laser transitions have been observed in optically pumped molecular iodine vapor by BYER et al. (1971). Selected vibrational rotational levels of the $(B^3\text{II}_{ou}^+ - X_g^1)$ electronic transition were excited by various visible lines emitted by a frequency-doubled Nd:YAG laser. Laser action on many more transitions was obtained subsequently (HÄNSCH, 1972), using a continuously tunable dye laser, repetitively pumped by a nitrogen laser. The total number of laser transitions in diatomic iodine molecules which can be excited by various visible pump frequencies is in the order of 10^6.

Narrowband tunable dye lasers also offer important advantages for the optical pumping of certain solid-state lasers. In particular it is possible to efficiently pump materials with rather narrow absorption lines which would otherwise utilize only a small fraction of a flashlamp continuum. VARSANYI (1971) has obtained strong laser action in the red from PrCl$_3$ and PrBr$_3$ crystals at room temperature, using a nitrogen

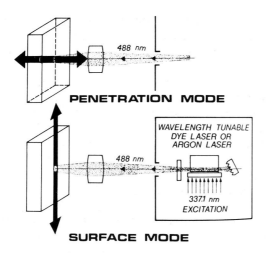

Fig. 5.7. Pumping of laser transitions in solid PrCl₃ with a pulsed tunable dye laser. Superradiant emission can be obtained in penetration modes and surface modes from a few ionic layers with dimensions of a few μm near the crystal surface. (VARSANYI, 1971)

laser-pumped wavelength-tunable dye laser as a pump source. The dye-laser light at 488 nm excites the upper laser level $^3P_0$ in the Pr^{3+} ion directly from the $^3H_4(2)$ ground state. Laser action terminates on the $^3F_2(2)$ state. These laser materials are unique because the active Pr^{3+} ions are not doped into a host crystal but are regular constituents of the lattice. The extremely high ion density, combined with a large quantum efficiency, provides an extremely large gain. Superradiant laser emission can be obtained from the first few ionic layers near the crystal surface, within linear dimensions of a few micrometers, and with excitation thresholds of less than a microjoule. Depending on the longest dimension of the active volume, laser emission occurs in a penetration mode or a surface mode, as illustrated in Fig. 5.7. VARSANYI has pointed out the potential of such miniature lasers for active integrated optical systems (see Section 5.10).

5.2.5. Photoionization

Excitation with tunable-dye-laser radiation can also be very useful for the selective photoionization of neutral atoms and molecules. The direct ionization from the ground state is prevented by the high ionization potential in most cases, but the presently available dye-laser wavelengths permit a stepwise two-photon ionization for many atoms

214   T. W. Hänsch

and some molecules. A photon $h\omega_1$ first excites a certain state of the discrete energy spectrum and a second photon $h\omega_2$ then ionizes the excited atom or molecule. This two-step ionization can be highly selective, and this is of particular interest for certain applications such as photochemistry. AMBARTZUMIAN and LETOKHOV (1972) have demonstrated the practical feasibility of this scheme with atomic rubidium vapor. The excited the $5^2P_{1/2}$ state of Rb with a ruby-pumped dye laser which provides pulses of 2 mJ energy with a bandwidth of about 0.2 nm at 794.76 nm. The ionizing light was generated by frequency doubling of part of the ruby laser beam. The ions, up to 10% of the theoretically estimated number limit, could be detected via the photocurrent between transverse electrodes in the Rb cell.

Photoionization from excited atomic states with tunable laser radiation is also finding interesting spectroscopic applications. STEBBINGS and DUNNING (1973) have irradiated a beam of excited, metastable argon atoms with ultraviolet light (295.0—307.0 nm) and have collected the produced ions with a particle multiplier. The tunable ultraviolet radiation was obtained by frequency doubling the output of a pulsed visible dye laser, pumped by a nitrogen laser, in an ADP crystal (see Section 5.5.1). The resultant UV beam had a linewidth of 0.05 nm, a peak power of several kW and a pulse length of 5 nsec. A plot of the ionization rate versus the light wavelength (Fig. 5.8) shows a number of sharp, discrete

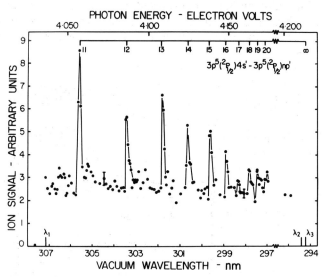

Fig. 5.8. Photoionization of a beam of metastable atomic argon with a frequency-doubled pulsed dye laser. The resonances in the ionization cross section correspond to autoionizing high Rydberg states. (STEBBINGS and DUNNING, 1973)

maxima, which are ascribed to autoionizing high Rydberg states of the configuration $3p^5(^2P_{1/2})np'$. The width of these resonance appears to be limited by the laser bandwidth.

5.2.6. Photodetachment

Pulsed tunable dye lasers of kilowatt peak power have also led to considerable progress in studies of the photodetachment of electrons from negative ions. LINEBERGER and his group (1972) have reported a series of beautiful experiments in which they measure the thresholds of photodetachment for a variety of ions with an energy resolution of better than 1 meV, i.e. 10—100 times more accurately than was previously possible with blackbody light sources. They have also, for the first time, been able to measure the energy dependence of the detachment cross sections near threshold and to verify theoretical predictions by WIGNER and others.

Figure 5.9 gives a schematic diagram of the photodetachment apparatus utilized (LINEBERGER, 1972). A mass-analyzed beam of negative ions travelling through a vacuum chamber is crossed by the light beam of a flashlamp-pumped dye laser. LINEBERGER used a grating-tuned laser with an output energy of 1—5 mJ per pulse, 300 nsec pulse length and 0.1—0.2 nm bandwidth, operating in the wavelength range 450.0 to 700.0 nm. The surviving ions are deflected into a Faraday cup, and the neutral detachment products are monitored with a gated particle detector. A small computer controls the dye-laser wavelength via a

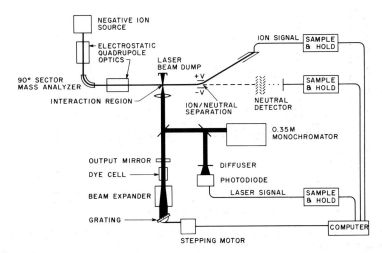

Fig. 5.9. Diagram of a tunable laser photodetachment apparatus. (LINEBERGER, 1972)

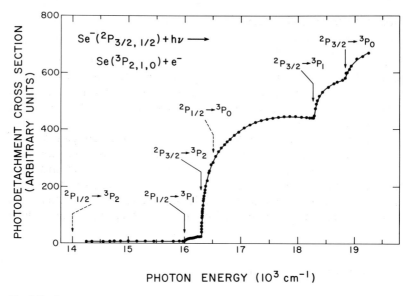

Fig. 5.10. Se⁻ photodetachment cross section versus dye-laser photon energy. The individual fine structure transition thresholds are labeled. (LINEBERGER, 1972)

stepping motor, monitors the various signals and computes relative photodetachment cross sections.

The first results were reported on the photodetachment of S⁻ ions (LINEBERGER and WOODWARD, 1970). Figure 5.10 shows the detachment cross sections for Se⁻ ions which were later obtained by HOTOP et al. (1973), using an improved version of the apparatus. Four different detachment thresholds, corresponding to different fine-structure levels of the Se⁻ ion and the Se atom, are clearly distinguishable. S⁻ and Se⁻ possess an outer p electron which leaves the atom after detachment in an s wave with zero orbital angular momentum. Wigner's threshold law predicts that the cross section will then be proportional to the square root of the excess energy and leads to an infinite derivative at threshold. The experimental confirmation is certainly impressive. A different threshold behavior could be studied with Au⁻, Pt⁻, and Ag⁻ ions (HOTOP and LINEBERGER, 1973).

The photodetachment of negative molecular ions is more difficult to interpret due to a large manifold of participating rotational and vibrational energy levels. Two-photon photodetachment with a near-resonant intermediate energy state can alleviate these difficulties, as demonstrated in measurements with C_2^- ions by LINEBERGER and PATTERSON (1972).

5.3. Nonlinear Spectroscopy

The high light intensities available from lasers have opened the way to various new nonlinear spectroscopic techniques involving changes of level populations, refractive indices, or other sample parameters by means of the irradiating lightfield which renders their response to the light amplitude nonlinear.

A number of such techniques have been developed through the use of fixed-frequency lasers of very limited tuning range. Among the most useful are the various methods of high-resolution saturation spectroscopy and Lamp-dip spectroscopy. The advent of narrowband widely tunable lasers is making these new and powerful techniques applicable to a wide variety of samples, and much exciting progress in spectroscopy may be anticipated.

5.3.1. Saturation Spectroscopy

The resolution of ordinary emission or absorption spectroscopy of gas resonance lines is limited by Doppler broadening owing to the random thermal motion of the gas particles, which often blurs important details of spectral line shape or structure. Laser saturation spectroscopy can virtually eliminate Doppler broadening: It utilizes the fact that a monochromatic laser wave interacts resonantly with atoms or molecules only over a narrow interval of the axial Maxwellian velocity distribution, i.e. with those atoms which are Doppler-shifted into resonance. If the laser intensity is sufficiently strong, it will induce saturation of the radiative transitions, causing velocity-selective level population changes which can be detected by a second probe wave.

Until recently such studies have been restricted to either gas-laser transitions or molecular absorption lines in accidental coincidence, because tunable dye lasers did not provide the required spectral resolution and stability.

The first application of a tunable dye laser for the high resolution saturation spectroscopy of atomic resonance lines was reported by HÄNSCH et al. (1971), who studied the yellow Na D resonance lines with a pulsed dye laser, repetitively pumped by a pulsed nitrogen laser. The linewidth of the dye laser was reduced from 300 MHz to about 7 MHz by means of an external confocal interferometer, serving as an ultra-narrow passband filter. This experiment demonstrated that it is now possible by means of wavelength-tunable dye lasers to apply the powerful techniques of saturation spectroscopy to virtually any atomic or molecular absorption line throughout the visible spectrum.

HÄNSCH et al. (1971) used a particularly sensitive and convenient method of saturation spectroscopy, previously used successfully in several saturation experiments with gas lasers and particularly well suited for use with a pulsed-laser source. A simplified scheme of the saturation spectrometer is shown in Fig. 5.11. The laser output is divided by a beam splitter into a weak probe beam and a stronger saturating beam, typically of a few mm in diameter, and these beams are sent in nearly opposite directions through a cell containing the absorbing gas sample. When the laser is tuned to the center of the Doppler profile of

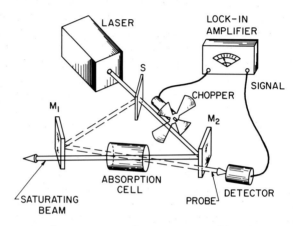

Fig. 5.11. Laser saturation spectrometer for spectroscopy without Doppler broadening. (HÄNSCH, 1972)

a given resonance line, both light waves can interact simultaneously with the same atoms, i.e. those with essentially zero axial velocity. The saturating beam then can bleach a path for the probe, reducing its absorption. In order to detect small bleaching effects, the saturating beam is periodically blocked by a rotating chopper, synchronized to the pulsed laser, and the resulting amplitude modulation of the probe is detected with a phase-sensitive amplifier.

In these experiments it proved possible to resolve the hyperfine splitting of both the Na ground state and the excited $^2P_{1/2}$ state, and to measure linewidths as narrow as 50 MHz, despite a Doppler width of 1350 MHz. The short laser-pulse length of some 30 nsec permitted measurements with a time delay between bleaching and probing, which reveals a remanent hole burning in the velocity distribution of the Na ground-state levels, due to a velocity-selective optical pumping cycle. This effect was utilized for the study of thermalizing eleastic collisions between Na and Ar buffer gas atoms.

HÄNSCH et al. (1972) used the same technique subsequently to study the red Balmer line H_α of atomic hydrogen at 656.3 nm, which has a particularly large Doppler width of almost 6000 MHz at room temperature. Doppler broadening normally masks all the finer details in the visible lines of the light hydrogen atoms. Our present accurate knowledge of the hydrogen fine structure is based almost entirely on radiofrequency spectroscopy and level-crossing techniques. Nonetheless, there have been countless efforts to study the visible Balmer lines with ever-increasing resolution, because an accurate determination of the wavelength of one of the line components is the basis for the determination of the Rydberg constant, one of the cornerstones in the evaluation of the fundamental constants of physics. But even when the Doppler width was reduced by using the heavier isotopes deuterium and tritium, and by cooling the emitting gas discharge to the temperature of liquid nitrogen or helium, it was not possible to isolate single fine-structure components, and it remained necessary to employ cumbersome deconvolution procedures for the determination of the wavelength.

Before any absorption could be observed on the red Balmer line, the hydrogen atoms had to be excited into the state $n = 2$. This was accomplished with a simple Wood discharge tube, operating at room temperature. The pulsed-laser source enabled the saturated absorption to be measured in the afterglow by quenching the discharge with an electronic switch about 1 μsec before the optical observation. In this way Stark broadening and level mixing due to electric fields in the plasma could be minimized. Figure 5.12 shows a saturation spectrum of H_α obtained in this way. The theoretical fine-structure components and their Doppler envelope at room temperature are given for comparison. The four strongest components are clearly resolved in the saturation spectrum, and the Lamb shift in hydrogen can, for the first time, be observed directly in the optical absorption spectrum. The linewidths of about 300 MHz are one order of magnitude narrower than the previous best results. The saturation spectrum includes one "crossover" signal, an artifact of the spectroscopic method, arising from two resonance lines with a common lower level. If the center of one of the line components can be located down to 4% of the width, the Rydberg constant could be determined to an accuracy of one part in $2 \cdot 10^8$, i.e. to the accuracy of the present standard of length.

The pulsed dye laser utilized offers a wide tuning range, relatively large peak powers up to several watts, and the opportuning for time-resolved studies; it also exhibits large amplitude fluctuations after spectral filtering (CURRY et al., 1973) and this limits its sensitivity for probe intensity changes to about 10^{-3}, when averaged over 100 laser pulses. Cw dye lasers with electronic intensity stabilization (HARTIG and

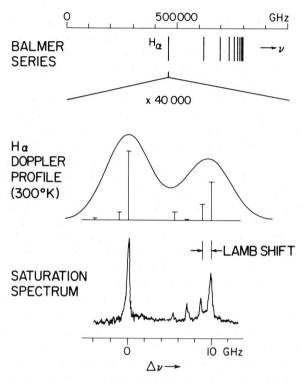

Fig. 5.12. Saturation spectrum of the red Balmer line of atomic hydrogen, with optically resolved Lamb shift. (HÄNSCH et al., 1972)

WALTER, 1973) should permit an increase in sensitivity by about three orders of magnitude. Sensitive saturation spectroscopy of dilute gases of very low absorption can also be performed with the method of inter-modulated fluorescence which has been demonstrated by SOREM and SCHAWLOW (1972) and analyzed by SHIMODA (1973). Another modification which is also applicable to the study of homogeneous spectral linewidths in solids is the observation of laser-induced fluorescence line narrowing, as described by SZABO and ERICKSON (1972).

5.3.2. Coherent Transient Effects

Short, intense light pulses, interacting with resonant transitions, can give rise to a number of interesting coherent transient effects, such as photon echoes or self-induced transparency, which often have analogies in phenomena observed in the microwave region in studies of nuclear magnetic resonance. Many workers have studied such effects with fixed-frequency ruby or CO_2 lasers. Tunable dye lasers can help to make coherent transient effects a useful tool for spectroscopic investigations, revealing such dynamic parameters as transverse relaxation due to phase-changing atomic collisions.

McCALL and HAHN (1967) were the first to demonstrate the phenomenon of self-induced transparency, a nonlinear optical propagation effect, using a ruby sample. A coherent light pulse, whose width is short compared to that of the relaxation time T_2 of a resonant optical medium, can pass through this medium as if it were transparent, provided its energy exceeds a critical threshold; below this critical energy the pulse is absorbed. For a qualitative explanation of this phenomenon, one can assume that the light pulse first loses energy in absorption, thereby adiabatically inverting the population of the resonant transition, and then regains this energy by stimulated emission. Naturally, this process can involve large time lags and a reshaping of the transmitted pulse. BRADLEY et al. (1970b) using a pulsed dye laser, pumped by a mode-locked ruby laser, reported observing self-induced transparency and dispersion delays in potassium vapor. The laser was tuned to the 769.9 nm absorption line of potassium and provided pulses of 5 MW peak power and 1 nsec duration. At low vapor pressure and a laser bandwidth of 0.05 cm^{-1}, the authors observed pulse delays up to 4 nsec at resonance, increasing with increasing light intensity, as would be expected from the theory of self-induced transparency. Large delays near resonance were observed at higher vapor pressure and were explained in terms of the regular linear dispersion of the medium.

BONCH-BRUEVICH et al. (1970) have observed other nonlinear phenomena in the passage of pulsed dye-laser radiation through atomic potassium vapor. A broadband laser, operating in the band from 760 to 800 nm with a peak intensity of 10^7 W/cm^2, was used in these experiments. Self-defocusing due to lenslike changes in refractive index was observed near resonance, and amplification was reported for a second, weak, slightly tilted laser beam, as well as the generation of a third coherent light beam; these phenomena were ascribed to stimulated four-photon scattering.

GRISCHKOWSKY and ARMSTRONG (1972) have reported a very clean and detailed study of self-defocusing of narrowband dye-laser radiation in dilute rubidium vapor. Dye-laser radiation on the low-frequency side of the $^2P_{1/2}$ resonance line (794.8 nm) was provided by a ruby-pumped dye laser, operating in a single resonator mode with less than 0.005 cm^{-1} bandwidth, 10 nsec pulse length, and several kW peak power. Complicating level degeneracies in the rubidium were removed by means of an 8 kG magnetic field. The observed self-defocusing was explained quantitatively by the electronic nonlinearity associated with "adiabatic follow-ing" of the laser field by the pseudomoments of the resonant atoms. In a subsequent publication, GRISCHKOWSKY (1973) reported the observation of group velocities as small as 1/14 of the velocity of light in vacuum. Intensity-dependent pulse velocities and a self-steepening of the pulses are predicted for higher intensities.

The reported experiments illustrate the exciting possibilities offered by studies of coherent transient effects in gases or solids with the help of pulsed tunable dye lasers. Much progress is certainly to be expected in the near future.

5.3.3. Nonlinear Susceptibilities

The nonlinear dielectric susceptibilities of solid crystals, which become observable with intense laser light and which have opened up the wide field of nonlinear optics, are also increasingly the subject of spectro-scopic studies.

When laser light is reflected from the surface of a metal or semi-conductor, a nonlinear susceptibility $\chi^{NL}_{(2\omega)}$ leads to the generation of weak, but observable radiation at the second harmonic frequency. PARSONS and CHANG (1971) measured the dispersion of the nonlinear suscept-ibility modulus $|\chi^{NL}_{(2\omega)}|$ for GaAs, InAs, and InSb within the fundamental wave range 0.73—1.0 μ, using a ruby-pumped dye laser of 0.5—2 MW peak power and 0.1 nm line width. Earlier measurements of this kind had been restricted to the discrete wavelengths available from fixed-fre-

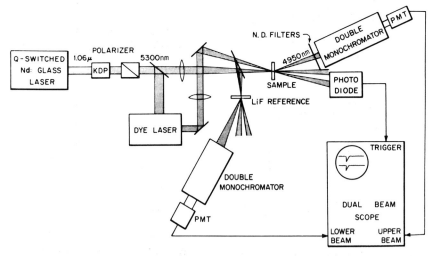

Fig. 5.13. Apparatus for resonant three-wave mixing in solids. (LEVENSON et al., 1972)

quency lasers. The sharp spectral structures observed are sometimes not correlated with the structures of the linear optical susceptibility and hence can provide additional information about electronic band properties. The same apparatus was used to determine the dispersion of three independent nonlinear optical susceptibilities in the hexagonal

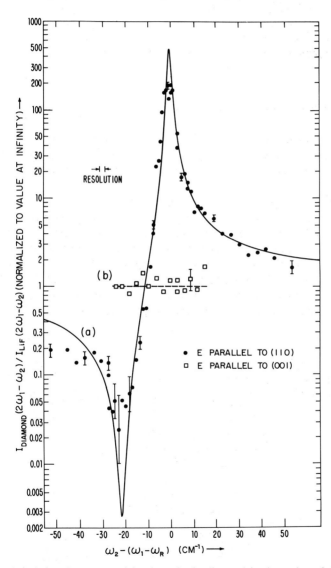

Fig. 5.14. Relative three-wave mixing intensity in diamond in the region of the optical phonon frequency. (LEVENSON et al., 1972)

II—IV semiconductors CdS and CdTe (PARSONS et al., 1972). The reported measurements are the first in the optical absorption region of these unaxial crystals; they will certainly be of much interest once the extensive theoretical calculations on these materials have been extended into this region.

A different nonlinear spectroscopic technique, coherent three-wave mixing, demonstrated by LEVENSON et al. (1972) and by WYNNE (1972), permits the measurement of third-order nonlinear susceptibilities of the type $\chi^{(3)}(-\omega_3, \omega_1, \omega_1 - \omega_2)$. Two tunable laser waves of angular frequency ω_1 and ω_2 are sent collinearly or with a slight angle into a transparent crystal. A third wave of angular frequency $\omega_3 = 2\omega_1 - \omega_2$ is generated in the crystal by coherent nonlinear mixing and the intensity of this wave is studied as a function of the frequency difference $\omega_1 - \omega_2$. Very dramatic resonant structures with intensity changes over more than four orders of magnitude are observed when the difference frequency approaches a Raman-active resonance of the crystal. The nonlinear susceptibility in this region is the sum of a real nonresonant electronic part and a complex Raman susceptibility, which induces a pronounced maximum and minimum of the intensity at the combination frequency. Since the Raman susceptibility can be determined from the linear Raman-scattering cross section, this technique can, for instance, be used to measure both the magnitude and the sign of the electronic nonlinear susceptibility without the uncertainties of other methods. LEVENSON and coworkers (1972) studied the effect in diamond, using the apparatus shown in Fig. 5.13. The two laser frequencies are provided by a frequency-doubled Nd : glass laser and by a tunable dye laser, pumped by part of the radiation of the first laser. The peak powers at the sample were 200 and 40 MW/cm^2, respectively, and the spectral resolution was 2 cm^{-1}. Figure 5.14 gives the normalized intensity generated at the new frequency as a function of the dye-laser frequency detuning. WYNNE (1972) investigated the same phenomenon in LiNbO$_3$, using two pulsed dye lasers, pumped simultaneously by a nitrogen gas laser and providing 1—5 kW peak power and 0.5 cm^{-1} bandwidth. The observed resonances are here due to transverse coupled photon-photon (polariton) modes.

5.3.4. Stimulated Light Scattering

The possibility of stimulated Raman scattering and stimulated Brillouin scattering was discovered soon after strong Q-switched ruby lasers became available. Stimulated light scattering differs from ordinary scattering in several respects. It occurs only above a distinct intensity threshold of the exciting beam. The scattered light is directional and can be highly monochromatic. And it is possible to obtain very high conversion efficiencies with strong laser pulses. Stimulated Raman scattering from liquids and gases has long been used as an efficient means to generate intense coherent radiation at new wavelengths, shifted from the original laser frequency by integer multiples of molecular vibration frequencies (ECKHARDT, 1966). Typical thresholds in bulk materials are on the order of 1—10 MW. Raman laser oscillation with a greatly reduced threshold of only 5 W has, however, been observed in a liquid fiber waveguide (IPPEN, 1970).

Excitation of stimulated Raman emission with tunable dye lasers could be useful to generate radiation at new wavelengths where appropriate dyes are difficult to find. Raman scattering in molecular hydrogen is of particular interest, because this substance provides the largest Raman shift of all known molecules (4155 cm^{-1}). SCHMIDT and APPT (1972) have succeeded in generating bright, blue-shiftet stimulated anti-Stokes light of first and second order from high-pressure hydrogen gas. The used a very powerful dye laser of 85 MW peak power which was excited by a Q-switched ruby laser. High-order multiple-stimulated anti-Stokes Raman scattering may ultimately provide a way to generate tunable coherent radiation in the vacuum ultraviolet, as indicated by experiments of MENNICKE (1971) who excited such scattering in solid hydrogen with a pulsed ruby laser.

Stimulated Raman scattering can also provide interesting spectroscopic information. BURRELL and KUNZE (1972b) have used a laser-pumped pulsed dye-laser of 10 kW peak power to excite stimulated Raman scattering from helium atoms in a cold plasma. The atoms were simultaneously interacting with a microwave field via two-photon absorption. The authors suggest that the observed effect may be useful for measuring high-frequency fields in plasmas.

Tunable dye-laser radiation can also be used to probe the gain of stimulated light scattering. REINHOLD and MAIER (1972) have used a ruby-pumped dye laser to measure stimulated Raman gain in ruby-pumped liquids in this way.

5.4. Other Spectroscopic Applications

The high spectral brightness of tunable dye-laser radiation can greatly increase the ease and effectiveness of a variety of other spectroscopic techniques. The most obvious examples are "classical" techniques which depend on the resonant excitation of atomic or molecular energy levels, such as the intensively utilized methods of optical-radiofrequency double resonance and level-crossing spectroscopy, or the related techniques of quantum beat spectroscopy. Other suggested new spectroscopic methods such as heterodyne spectroscopy, depend critically on the coherence of laser light. The advent of tunable lasers promises to turn them into viable tools.

5.4.1. Level-Crossing Spectroscopy

Characteristic changes in the polarization and angular distribution of fluorescent light from atomic gases can be observed if atomic energy levels are forced to cross by changing an applied magnetic field. These level-crossing phenomena can be explained in terms of the interference of different radiating channels in the light-emitting atom. Level-crossing spectroscopy with conventional light sources or electron impact excitation has been used extensively to study the fine and hyperfine structure of spectral lines, which is normally blurred by Doppler broadening, or to measure g-factors and lifetimes. The observed line-widths are normally determined by the radiative lifetimes of the crossing levels. It has, however, been demonstrated that linewidths narrower than the natural linewidth can be observed when the atoms are excited with a short pulse and the fluorescence is measured after a time lag, so that atoms with a longer than normal lifetime contribute predominantly to the signal.

Delayed level-crossing experiments are very difficult when the atoms are excited with pulsed discharge lamps, but a pulsed dye laser can excite a sufficient number of atoms to permit measurements after a time lag on the order of 5—10 times the natural lifetime. ERDMANN et al. (1972) have reported preliminary experiments with the $3^2P_{3/2}$ level of atomic sodium, using a flashlamp-pumped dye laser with electronic shut-off via an internal Lyot filter (see Section 5.1.7). They were able to determine the hyperfine splitting of the studied level with a higher precision than in earlier experiments. SCHENCK et al. (1973) have performed a similar study of the yellow sodium line, using a pulsed dye laser pumped by a nitrogen laser.

5.4.2. Quantum-Beat Spectroscopy

Excitation with short pulses can prepare atoms in a coherent super-position of closely spaced, but non-overlapping states. Interference between the different decay channels to a common ground state can then lead to modulations in the fluorescence light at frequencies correspond-ing to the energy splitting between the states. Such "quantum beats" have been studied in numerous experiments, using optical excitation, electron impact, or beam–foil excitation. They can be used to study level splittings without masking Doppler broadening, and without perturbation of the atom by external fields. Optical excitation provides high selectivity, but measurements with conventional spectral lamps and Kerr cell shutters have been limited to relatively low modulation frequencies (a few MHz), and they require lengthy photon-counting due to very low intensities.

Pulsed tunable dye lasers can excite atomic quantum beats with a very impressive signal-to-noise ratio, as first demonstrated by GORNIK et al. (1972). These authors excited the triplet resonance line in an ytterbium atomic beam with a nitrogen-laser-pumped dye laser. The pulse length of 10 nsec is short compared with the 860 nsec lifetime of the excited state. A static magnetic field of 1 Gauss split the $m = \pm 1$ sublevels of this state by a few MHz. For the even isotopes, a 100% modulation of the fluorescence light could be observed with a single laser pulse.

HAROCHE et al. (1973) have been able to observe hyperfine quantum beats of much higher frequencies in the blue, second-resonance line of atomic cesium ($6^2S_{1/2} - 7^2P_{3/2}$). They irradiated cesium vapor with the linearly polarized light of a dye laser, pumped by a pulsed nitrogen laser, with a pulse length of about 2 nsec. The upper cesium state has a lifetime of 135 nsec and is split into four hyperfine components, spread over about 200 MHz. The laser bandwidth is sufficient for the simultaneous excitation of the three hyperfine levels which can be reached from a given ground-state level. The modulated fluorescence light is detected with a fast photomultiplier, followed by a sampling oscilloscope and a digital signal averager. The quantum beat signals for the line starting in the ground-state level $F = 3$ are displayed in Fig. 5.15.

EXCITATION FROM THE F=3
GROUND STATE

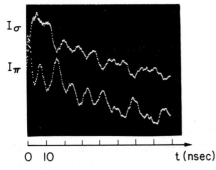

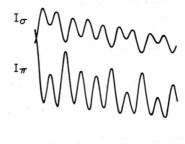

Fig. 5.15. Quantum beats in spontanious emission. Intensity modulation of the Cs fluorescence line $7^2P_{3/2} - 6^2S_{1/2}$ (F = 3) after excitation with a short dye laser pulse, due to beats between three hyperfine components. Upper and lower trace correspond to the polarizations I_π and I_σ respectively. Top: experimental results, bottom: theoretical prediction. (HAROCHE et al., 1973)

The upper and lower traces were recorded at perpendicular and parallel linear polarization, respectively. The agreement with the calculated modulation curves below is excellent. By Fourier analysis of the beat signals HAROCHE and coworkers were able to reproduce the known hyperfine intervals within $\pm 1.5\,\text{MHz}$.

5.4.3. Heterodyne Spectroscopy

Narrowband cw dy lasers promise an important breakthrough in optical heterodyne spectroscopy, even though no experimental demonstration has yet been reported. Several authors have discussed the possibility of using a laser as a local oscillator and mixing the coherent laser light with the light of an incoherent source in a fast photodetector with normal quadratic amplitude characteristic. This mixing will result, apart from self-mixing terms of the two sources, in a radiofrequency noise signal which is a faithful copy of the information present in the source at optical wavelengths near the laser frequency. By electronic signal processing it is possible to simulate a spectral filter characteristic of very narrow bandwidth and almost arbitrary shape.

BENEDEK (1969) has discussed some possible applications of optical mixing spectroscopy in physics, chemistry, biology and engineering. At optical frequencies, heterodyne spectroscopy does not offer the large advantage in sensitivity which we are accustomed to in the radio-frequency region. But for point-like sources and with a detector of high quantum efficiency the signal-to-noise ratio could be comparable to that of other spectroscopic methods. NIEUWENHUIJZEN (1970) has experimentally demonstrated the feasibility of optical heterodyne spectroscopy of incoherent light emitted by a gas discharge, and of starlight collected by a telescope, using a He–Ne gas laser as local oscillator. He discusses possible applications for stellar spectroscopy and stellar interferometry. HINKLEY and KELLEY (1971) have proposed that heterodyne spectroscopy in the infrared could be utilized to observe emission lines from pollutants in industrial smoke stacks.

5.5. Generation of Tunable UV and IR by Nonlinear Frequency Mixing

The wavelength range of presently available laser dyes is limited to the region from 340—1200 nm. Tunable coherent radiation at new wavelengths can be generated, however, by frequency mixing of dye-laser radiation in nonlinear optical materials. Tunable ultraviolet radiation down to 216 nm has been obtained in this way by harmonic and sum frequency generation in nonlinear crystals. Experiments to generate tunable radiation in the near and middle infrared as the difference frequency of two laser oscillators have also been successful. Optical parametric oscillators offer a different alternative for the generation of tunable infrared radiation. The tuning speed and convenience of such devices can be improved by using a wavelength-tunable dye laser as pump source.

The fundamentals of the vast and rapidly growing field of nonlinear optics have been summarized in thorough reviews by BLOEMBERGEN (1965) and by TERHUNE and MAKER (1968). SINGH (1971) has complied a survey of the materials presently used. The sum or difference frequency of two laser waves can be generated in crystals of the non-centro-symmetric group, where the dielectric polarization exhibits a term quadratic in the inducing field. In order to permit the build-up of the signal over large "coherence lengths" by constructive interference, it is necessary that the momentum vectors of the interacting light waves satisfy the same additive relationship as the frequencies. This phase-matching condition can be rather restrictive in practice. It can be satisfied, despite the dispersion of the refractive index, with certain birefringent materials, by chosing the correct angular direction and polarization of the incident light waves and/or by accurate control of the crystal temperature.

5.5.1. Harmonic and Sum Frequency Generation

The generation of tunable ultraviolet radiation by frequency doubling of visible dye laser light is closely related to the well-established procedure of second harmonic generation with fixed frequency lasers, and has been demonstrated in many laboratories. The doubling crystal can be conveniently placed outside the dye-laser cavity if peak powers in the kW to MW range are available. Since phase matching requires different crystal orientations for different wavelengths, it is generally necessary to rotate the crystal, when the dye laser wavelength is tuned.

At lower laser intensities, the power of the second harmonic radiation grows with the square of the input intensity, i.e. the energy conversion efficiency is proportional to the input intensity. At high intensities the efficiency levels off at an approximately constant value between 10 and 40%, partly determined by nonlinear absorption processes in the crystal. The dye-laser light is generally focused into the doubling crystal to obtain high intensities. Too tight focusing reduces the efficiency, however, because the angular spread of the beam increases and simultaneous phase matching for the different angular directions is

difficult to achieve. For a 38 mm-long crystal of KDP (potassium dihydrogen phosphate) cut under $\theta = 64.4°$, for instance, the angular
divergence of incident monochromatic laser radiation should not
exceed 2' of arc or 0.66 mrad for efficient second harmonic generation,
according to calculations by Kuhl and Spitschan (1972). To minimize
the beam divergence for a given cross section, it is very desirable that the
dye laser oscillates in the fundamental TEM_{00} mode. The requirement of
small beam divergence can be somewhat relaxed, if the dye laser
operates in a wider spectral band of about one nm width, because the
phase-matching condition can then be satisfied for the different directions by the generation of the sum frequency from two different parts
of the dye laser spectrum.

Bradley et al. (1971) have used an ADP (ammonium dihydrogen
phosphate) crystal to double the output of an interferometrically narrowed rhodamine 6 G laser, side-pumped by a frequency doubled Nd: glass
laser. They obtained pulses of 20 mJ and megawatt peak power, continuously tunable from 280 to 290 nm, with a bandwidth of 0.2 nm. The
reported conversion efficiency was 10%. Hamadani and Magyar (1971)
obtained ultraviolet pulses of similar energy, tunable between 340 and
400 nm and width a bandwidth of 5 pm, by doubling the output of a
ruby-pumped dye laser in a $LiIO_3$ crystal. Dunning et al. (1972) have
generated ultraviolet pulses, continuously tunable from 250 to 325 nm,
by frequency doubling the output of a nitrogen laser pumped dye laser.
A 25 mm-long cooled ADP crystal (60° z-cut) yields 11% conversion
efficiency at a dye-laser peak power of 60 kW. The dye-laser linewidth is
narrowed to less than 0.2 nm by means of a diffraction grating and a
beam-expanding prism inside the cavity. The authors used a beam con-
vergence angle of 3.2 mrad inside the crystal and discuss the parameters
affecting the conversion efficiency. Figure 5.16 shows the fundamental
power and doubled output power for different dye solutions as a function of wavelength. More recently, Dunning et al. (1973) have been able
to generate tunable radiation from 300 down to 230 nm, using a 10 mm-
long doubling crystal of lithium formate monohydrate at room temperature. The conversion efficiency reached 2% at fundamental powers in
excess of 50 kW.

Other workers have successfully doubled the frequency of flashlamp-
pumped dye lasers with output powers in the kW range. Jennings and
Varga (1971) have generated a peak power in the order of 10 W in
0.6 μsec-long pulses, tunable from 250 to 290 nm, by using a 90° phase-
matched temperature-tuned ADP crystal outside the dye-laser cavity.
Kuhl and Spitschan (1972) doubled the output of a flashlamp-pumped
rhodamine 6G laser with a 38 mm-long KDP crystal and have achieved
power conversion efficiencies of 10—18% for laser output powers as low

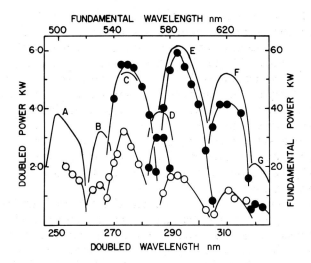

Fig. 5.16. Frequency doubling of a nitrogen-laser-pumped dye laser in ADP.—fundamental dye laser input power, -•-doubled output power, 60° z-cut crystal, 2.5 cm long, -○-doubled output power, 90° z-cut crystal, 1 cm long. A dye mixture of 7-diethylamino-4-methylcoumarin (7D4MC) and acriflavine in ethanol, B 7D4MC and fluorescein disodium salt (FDS) in methanol, C FDS in methanol, D FDS and rhodamine 6G (R6G) in ethanol, E R6G in ethanol, F rhodamine B (RB) in ethanol, G RB and cresyl violet acetate in ethanol. (DUNNING et al., 1972).

as 15—25 kW. The tuning range extended from 285 to 310 nm. The laser bandwidth could be varied between 8 nm and 1 pm by means of intracavity interference filters and Fabry-Perot interferometers, and oscillation could be restricted to the TEM_{00} mode. The influence of the spatial and spectral characteristics of the laser beam on the conversion efficiency into the ultraviolet were studied in detail.

It is also possible to generate narrowband tunable ultraviolet radiation from the output of a broadband dye laser by phase-matched symmetric sum frequency generation in a nonlinear crystal. This scheme was demonstrated by JOHNSON and SWAGEL (1971), who used a rhodamine 6G dye laser, side-pumped by the green light of a frequency-doubled Nd:glass laser of 2 MW power, and a external KDP crystal. Wavelength tuning is possible by simply rotating the crystal. Similar results, with a conversion efficiency of 3%, were also reported by KUHL and SPITSCHAN (1972), using their 28 kW flashlamp-pumped dye laser. A disadvantage of this method is the relatively large and somewhat uncontrolled UV linewidth, typically on the order of about one nm.

Intracavity frequency doubling is possible with cw dye lasers and other dye lasers of low power and relatively long pulse width. A

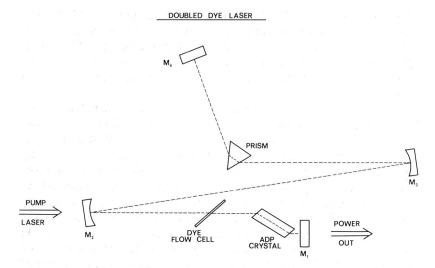

Fig. 5.17. Pulsed dye laser with frequency-doubling ADP crystal inside the four mirror cavity. (WALLACE, 1971)

commercial tunable UV laser, pumped by a repetitively pulsed, frequency-doubled Nd:YAG laser, uses an angle-tuned ADP crystal inside the dye laser resonator, as shown in Fig. 5.17 (WALLACE, 1971). The output (up to 200 W peak power, 5 mW average power, 0.2—0.4 μsec pulse length, and 2 cm^{-1} bandwidth) can be tuned between 261 and 315 nm. GABEL and HERCHER (1972) have employed an intracavity crystal of lithium formate in a cw laser to generate continuous tunable ultraviolet in the region 290 to 315 nm.

The highest second harmonic frequencies have so far been obtained with lithium formate, which becomes opaque below 230 nm. Although KDP and ADP are transparent down to 200 nm, they do not permit the generation of wavelengths below 250 nm by frequency doubling in an efficient 90° phase-matched configuration. Ultraviolet radiation at shorter wavelengths can be obtained, however, as the sum frequency of two different laser waves of different frequencies, as demonstrated by AKH-MONOV et al. (1969), who generated 212 nm as the sum frequency of the infrared light (1.06 μm) of a Nd laser and its fourth harmonic in the ultraviolet. The use of dye lasers should permit the generation of tunable radiation in this way.

DINEV et al. (1972) have produced tunable ultraviolet radiation in the range from 216 to 234 nm by threefold sum frequency generation in a calcite crystal, using the third-order nonlinear susceptibility of this material. They mixed the output of a Q-switched Nd:glass laser with two-stage amplifier (70 MW) with its second harmonic (20 MW) and with the output of a broadband rhodamine 6G or rhodamine B dye laser (2—4 MW), pumped by part of the second harmonic light. The output peak powers reached 15 kW, and wavelength tuning was possible by rotating the crystal. The strong dispersion of the calcite results in a linewidth as narrow as 15 pm.

Very promising for the generation of tunable radiation in the vacuum ultraviolet, and possible even in the soft x-ray region, is the third harmonic generation and threefold sum frequency generation is gases, such as Rb or Cs vapor, in which phase matching can be achieved by compensating an anomalous dispersion with an appropriate buffer gas, such as He. This scheme has been proposed and demonstrated by HARRIS and coworkers and wavelengths down to 110 nm have already been generated, using as the primary source a nontunable picosecond Nd laser with frequency doubler and quadrupler (YOUNG et al., 1971; MILES and HARRIS, 1972). Coherent radiation at 118.2 nm could also be produced by third harmonic generation in a phase-matched mixture of negatively dispersive Xe and Ar buffer gas (KUNG et al., 1973). A conversion efficiency of 2.8% has been obtained at an input power of 13 MW, and the mixture of Xe and Ar appears suitable for generation from 354.7 to 118.2 nm. The required primary powers are within the reach of presently available high-power tunable dye lasers.

5.5.2. Difference Frequency Generation

Considerably less work has so far been reported toward the generation of infrared radiation by nonlinear mixing of dye-laser light, even though generation of very narrowband and reliably tunable infrared light should be possible over a wide wavelength range and is of considerable practical interest, for instance for molecular spectroscopy or pollution monitoring.

DEWEY and HOCKER (1971) have produced coherent infrared light between 3 and 4 μm in angle-tuned $LiNbO_3$ crystal as the difference frequency of a ruby laser and a ruby-pumped DTTC iodide dye laser. Peak powers of several kW and a bandwidth of 3—5 cm^{-1} were reported, but the $LiNbO_3$ crystal restricts generation to wavelengths lower than 4.5 μm. HANNA et al. (1971) have obtained tunable radiation in the middle infrared between 10.1 and 12.7 μm in a similar fashion, using a proustite crystal. With a 290 kW ruby laser and a cryptocyanine dye laser they produced peak powers on the order of 100 mW. DECKER and TITTEL (1973) have generated kilowatt power in the wavelength range of 3.2—5.6 μm in proustite, using a Q-switched ruby laser of 900 kW and a ruby-pumped dye laser of 20—40 kW power, operating between 790 and 890 nm. Appropriately oriented proustite crystal should theoretically permit coverage of the full range from 2.5 to 13 μm. MELTZER and GOLD-BERG (1972) produced infrared radiation tunable from 4.1 to 5.2 μm by mixing a DTTC iodide dye laser with its ruby pump in a lithium iodate crystal internal to the dye laser cavity. They obtained 100 W peak power from a 4 MW ruby laser. Lithium niobate can again be utilized for generation of radiation in the far infrared, as demonstrated by FARIES et al. (1971), who produced submillimeter waves between 20 and 38 cm^{-1} by mixing the outputs of two ruby lasers, operating at different frequencies.

SOROKIN and coworkers (1973) have recently demonstrated and interesting scheme for the generation of narrowband-tunable infrared radiation, using a parametric four-wave mixing process in alkali vapor. Figure 5.18 gives a scheme of the mixing process in potassium. Two pulsed dye lasers provide two input beams at frequencies ν_L and ν_P. The

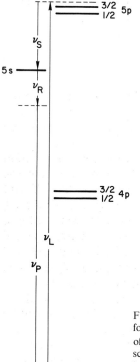

Fig. 5.18. Generation of tunable infrared by parametric four wave mixing in K vapor. Tunable dye laser radiation of frequency ν_L excites stimulated electronic Raman scattering at frequency ν_S. Mixing with a second dye laser wave of frequency ν_P results in the phase-matched generation of infrared light with frequency ν_R. (SOROKIN et al., 1973)

laser frequency v_L is close to the blue resonance line $4s$—$5p$ and excites stimulated electronic Raman emission at a frequency v_S to the state $5s$. Mixing with the tunable frequency v_P of the second laser results in the generation of an infrared beam of frequency v_R. (Under certain circumstances the parametric four-wave mixing could also be observed without the second dye laser.) Phase matching can be achieved by proper adjustment of the frequency v_L, without the need for any buffer gas, as employed in the phase-matched threefold sum frequency generation in alkali vapor by HARRIS and coworkers (YOUNG et al., 1971; MILES and HARRIS, 1972).

In the reported experiment, the dye lasers were repetitively pumped by a nitrogen laser and provided peak powers of about 1 kW at a bandwidth of 0.1 cm^{-1}. The alkali vapor was contained in a heat-pipe oven of 30 cm active length at a pressure between 0.5 and 20 Torr. With potassium it was possible to generate wavelengths between 2 and 4 μm with peak powers of about 0.1 W. Rubidium permitted a continuous tuning range from 2.9 to 5.4 μm. In principle, using other resonances, this method could be used to generate tunable radiation out to much longer wavelengths.

5.5.3. Pumping of Optical Parametric Oscillators

WALLACE (1972) has generated rapidly tunable infrared radiation in the region from 0.72 to 2.6 μm by pumping an optical parametric oscillator with a tunable dye laser. Parametric oscillators are reactive devices, closely related to parametric amplifiers and oscillators in the microwave region, and in effect reverse the process of sum frequency generation in a nonlinear crystal (HARRIS, 1969; BYER, 1973). Optical parametric oscillators are of considerable practical interest for the efficient generation of tunable coherent radiation in the near and middle infrared, where laser dyes are not available. WALLACE's converter consisted of a 5 cm-long LiNbO$_3$ crystal in a temperature-controlled oven, placed between two resonator mirrors which are highly reflective in the infrared and transparent at shorter wavelengths. Laser pump light enters through one end mirror. If the parametric gain exceeds the losses, oscillation builds up on two different frequencies, the signal and the idler, whose frequencies add up to the pump frequency and which are mutually phase-matched for difference-frequency generation with the pump. The wavelengths of signal and idler can be tuned by changing the phase-matching conditions. Usually, with a fixed-frequency-laser pump source, this requires a change in the crystal temperature. Since the temperature gradient over the crystal has to be very small, temperature

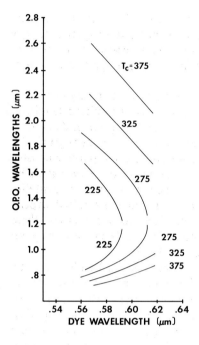

Fig. 5.19. Rapidly tunable parametric oscillator, pumped by a pulsed dye laser. Oscillator signal and idler wavelength versus dye laser wavelength for different temperatures of the LiNbO$_3$ crystal. (WALLACE, 1972)

tuning results in very slow tuning rates. Dye-laser pumping, on the other hand, permits rapid tuning at a fixed crystal temperature by changing the pump wavelength. WALLACE used as pump source a rhodamine 6G dye laser, longitudinally pumped by the second harmonic of a Q-switched Nd:YAG laser. It provided pulses of 100 μJ energy, tunable from 560 to 620 nm, with a bandwidth of 1 cm^{-1}. The conversion efficiency in the parametric oscillator reached 25%. Typical tuning curves are shown in Fig. 5.19.

5.6. Photochemistry, Isotope Separation

Photochemistry, i.e. the chemistry of optically excited atoms and molecules, currently offers some of the most promising opportunities for the application of wavelength-tunable lasers. Not only can differences in the optical absorption spectra of materials be utilized for the spectroscopic analysis, but it is also possible to bring about selective chemical or physical changes in matter by irradiation with light. The high selectivity of the optical excitation in the gas phase is suitable for the photochemical separation of different isotopic species, and the selective excitation with intense laser light should permit the control of specific chemical reactions and might even make it possible to synthesize new chemical compounds which are not otherwise accessible. These prospects have attracted the attention of many workers and could eventually have a large economic impact.

Photochemical processes have long been known, but only in the last fifteen years has there been a major surge of interest in photochemistry, due partly to improved analytical techniques and progress in theoretical understanding. An important reference work is the monograph by CALVERT and PITTS (1966). The possible uses of lasers in photochemistry have been discussed in recent interesting and detailed review papers by LETOKHOV (1973) and by MOORE (1973). Several successful feasibility studies have already been reported, and though these initial experiments have mostly been carried out with fixed-frequency lasers, there can be little doubt that wavelength-tunable dye lasers will play an important role in the future development of this field.

A number of promising reaction mechanisms can be illustrated by the methods for optical isotope separation which have been discussed so far. Photochemical isotope separation promises high separation factors, high efficiencies and much reduced costs; it could open up new applications for isotopes in research and technology. It is well known that atomic energy levels exhibit isotope shifts due to hyperfine interactions, nuclear volume effects, and, for lighter atoms, changes in the reduced mass. Nuclear-mass changes are also responsible for isotope shifts in the vibrational and rotational frequencies of molecules. The absorption lines can be displayed by as much as several cm^{-1} for different isotopic species, i.e. the isotope shifts can be considerably larger than the Doppler width of gaseous absorption lines. It is consequently possible to selectively excite a single species in an isotopic mixture by irradiation with monochromatic light of the right wavelength. Competing thermal excitation can be a problem for low rotational and vibrational molecular levels. However, the wavelengths provided by tunable dye lasers are sufficiently short to reach electronically excited states of some molecules, and this is in general advantageous for the selectivity of the subsequent processes.

5.6.1. Fast Photochemical Reactions

The main problem in laser isotope separation is the removal of the selectively excited species by chemical or physical means, before it decays back into the ground state, or before competing processes like collisional excitation transfer can interfere.

One solution is to induce a fast chemical reaction of the excited atom or molecule A^* with an appropriate reaction partner, forming one or more chemical compounds which can be removed chemically. The selection of the acceptor R is complicated by the requirement that the reaction rate with partner A in the ground state must be sufficiently slow. Following an initial success by Kuhn and Martin in 1933, this scheme of photochemical catalysis has nonetheless been successfully applied by a number of workers to the separation of isotopes of Hg, Cl, O, and C, using excitation with conventional spectral lamps (LETOKHOV, 1973). An attempt by TIFFANY et al. (1968) to separate Br_2 by selective excitation with 694.3 nm ruby-laser radiation was unsuccessful, because the subsequent Br_2^* + olefin collisions yielded not molecular products but Br atoms, which then initiated the exchange of isotopes through free-radical chemistry. The first apparently successful isotope separation by selective laser excitation was reported by MAYER et al. (1970). They used an HF laser near 2.7 μm to excite an OH stretching vibration in a mixture of CH_3OH, CD_3OD, and Br_2. The excited CH_3OH molecules reacted with the bromine, leaving CD_3OD as the remaining alcohol. The flexibility offered by tunable-dye-laser radiation will certainly open up many new possibilities for the photochemical separation of atoms and molecules.

5.6.2. Two-Step Photoionization

Another possibility for the removal of the excited species A^* is its subsequent photoionization with strong ultraviolet light. This resonant two-step ionization, as discussed in Section 5.2.5, takes full advantage of the selectivity of resonant excitation, in contrast to direct photoionization from the ground state. The ions could, for instance, be removed by deflecting fields in a molecular-beam apparatus. This technique appears feasible for the technically important separation of the uranium isotopes U^{235} and U^{238}. They exhibit an isotope shift of 1.4 cm^{-1} in the absorption line at 402.4 nm which can easily be reached with pulsed dye lasers (MOORE, 1973).

5.6.3. Two-Step Photodissociation

A more chemical method is the photodissociation of selectively excited molecules A^* into stable compounds or fragments which can be trapped chemically. Resonant two-step dissociation with laser light was reported by AMBARTZUMIAN and LETOKHOV (1972), who excited the vibrational state $v = 3$ in HCl with the output of a Nd : glass laser and dissociated the excited HCl molecules with the fourth harmonic of this laser at 265 nm. The product Cl atoms were chemically trapped by added NO, forming NOCl. This technique should be rather generally applicable if a tunable dye laser is used for the resonant excitation. A fixed-frequency laser could be used for the dissociating UV radiation in many circumstances.

5.6.4. Photopredissociation

A simpler, though less general, method of isotope separation is the photopredissociation of molecules by resonant excitation from the ground state. Molecules with "crossing" electronic states can have excited energy levels which are quasi-bound (metastable) with respect to dissociation and which can be reached via relatively sharp absorption lines. Selective optical excitation then can directly yield isotopically enriched dissociation products. YEUNG and MOORE (1972) have demonstrated such an isotope separation in practice. They used a doubled ruby laser, 347.2 nm, to photolyze an equimolar mixture of H_2CO and D_2CO. D_2 was produced in a sixfold abundance over H_2. Much better separation factors should be obtainable with tunable dye lasers which make it possible to choose a spectral region with widely different absorption coefficients for the two isotopic species. MOORE (1973) has suggested that this particular predissociation process could be applied to enhance the sensitivity of C^{14} dating by orders of magnitude.

5.7. Dye-Laser Applications in Biology

It has long been known that light can influence living tissue and biological processes. Photosynthesis and vision are perhaps the most dramatic examples of such interactions. Ultraviolet photobiology has been a vital research field for more than 30 years. It is therefore not surprising that applications of laser light in biology were explored as soon as the first fixed-frequency lasers became available. The effects of laser light on living cells can be ascribed to a number of primary processes, in particular thermochemical changes due to transient heating, photochemical reactions, vaporization of protoplasma and, for very intense pulses, acoustical and mechanical damage. These interactions have been used by numerous workers to study a variety of biological phenomena. Many biological applications of fixed-frequency lasers are reviewed in the monographs by WOLBARSHT(1971) and by GOLDMAN and ROCKWELL(1971). They include the microsurgery of organelles, in particular the selective destruction of parts of the cell walls, nucleus, cytoplasm, mitochondria or chloroplasts by the use of a laser attached to a microscope. Laser light has also been employed as a microprobe for spectroscopic emission analysis of biological microsamples. Lasers have permitted interesting studies of cellular metabolism, and laser-induced photochemical reactions are of particular interest for macromolecular studies in biochemistry. Numerous experiments with amino acids, proteins and nucleic acids have already been reported. The modification of DNA by irradiation with ultraviolet light, perhaps via thymine dimerization, could find practical applications for the selective generation of genetic mutations.

Since photochemical reactions, as well as simple heating by light absorption, depend critically on the light wavelength, tunable dye lasers hold promise of a great increase in the potential of laser light as a tool of biological research. Dye-laser light can be matched to selected absorption maxima of biological materials. Sometimes it is necessary to increase the light absorption by adding certain dyes. Tunable lasers will offer a wider choice of such dyes, thus permitting lower concentrations, a factor which can be important for investigations in vivo.

5.7.1. Photolysis of Biological Molecules: Reaction Kinetics

Pulsed dye lasers have already been applied by several researchers to photodissociate specific ligands from large biological molecules in studies of biochemical reaction kinetics. Flash photolysis is a well-known and widely used technique for chemical kinetic investigations. Molecules in a sample solution are dissociated by an intense light pulse, normally provided by a flashlamp. This photolysis and any subsequent chemical reactions of the dissociation products can be monitored by observing the optical absorption spectrum of the sample at right angles by means of a continuum light source, spectrometer and fast photodetector. Dye lasers offer some obvious advantages over flashlamps in this technique. The tunable wavelength permits better control over the photochemical dissociation and the high intensity obtainable in small volumes can provide a high degree of photolysis, even if absorption coefficient and quantum efficiency of dissociation are low.

MCCRAY(1972) has reported interesting studies of the oxygen recombination kinetics of various hemoglobin molecules, the oxygen carriers of blood cells. He dissociated oxyhemoglobin in a small cuvette $(5 \times 5 \text{ mm}^2)$ with the light of a flashlamp-pumped rhodamine 6G dye laser. The dissociation of the bound oxygen is highly specific, but requires very high light intensities. 60—70% of the oxyhemoglobin was photolyzed by 1 J dye-laser pulses at 585nm. The recombination with dissolved oxygen results in characteristic absorption changes in the

436—456 nm region. The reaction is rapid, with rate constants between 0.1 and 1 msec, and would cause severe problems in more conventional kinetic studies, such as stop-flow mixing. The reported measurements reveal the existence of two different hemoglobin components whose reaction rates differ by a factor of 7.

CHANCE and ERECINSKA(1971) have utilized a similar technique to investigate the oxidation of the enzyme cytochrome a_3 in mitochondrial membranes. Cytochrome a_3, or oxidase, is the last member in a chain of cytochrome enzymes responsible for electron transport in the synthesis of energy-storing ATP molecules. The mechanism of this important process is still largely unknown, but the various cytochromes are known to exhibit in the blue parts of the spectrum characteristic absorption bands which can be utilized to monitor their chemical status. Those investigators dissociated CO from CO-liganded cytochrome a_3 in a flow cell by irradiation with a 100 mJ pulse from a flashlamp-pumped rhodamine 6G dye laser. The resulting reduced cytochrome a_3 absorbs monitoring light at 445 nm until it is oxidized by admixed dissolved oxygen. The measured rate constant for this process at an O_2 concentration of 17 M/l was about 1 msec. The effect of uncoupling agents, in particular ATP, was also studied. Later measurements at subzero temperatures (ERECINSKA and CHANCE, 1972) revealed the probable existence of an intermediate product of the oxidase-oxygen reaction with an absorption band at 428 nm.

DUTTON et al.(1971) irradiated cooled chromatophores of chromatium D with the light of a ruby laser (694 nm) and a ruby-pumped dye laser (868 nm). They observed spectral changes due to the photo-oxidation of cytochrome c by the bacteriochlorophyll. The measurements suggest the existence of two modes of electron transfer with different time constants. The reactions did not depend on the exciting wavelength, however.

5.8. Applications in Medicine

Some applications of dye lasers in medicine can readily be envisioned, but no practical applications have yet been reported. One may judge their future potential from the wealth of literature on medical uses of fixed-frequency lasers. Many references are given in the recent monograph on lasers in medicine by GOLDMAN and ROCKWELL(1971) and in the first published volume of a review series, edited by WOLBARSHT(1971).

Certain applications, such as the attempted destruction of cancers and tumors with laser light, the removal of dental caries, or the treatment of fractured bones, have met only rather moderate success;

others appear more promising. A tightly focused intense laser beam can be used for bloodless and sterile surgery, for instance. Intense laser pulses can remove tattooing, without pain and without the scarring encountered in surgical removal. Laser irradiation has also been helpful in the treatment of certain disfiguring skin pigmentations, such as the congenital purple-reddish „portwine" marks. The most successful, advanced and widely used application of lasers in medicine is certainly eye surgery by photocoagulation. An intense laser beam focused into the fundus causes local heating with formation of scar tissue, which can prevent detachment of the retina following a retinal hole or tear. The laser can also be used to coagulate blood vessels in order to obstruct the flow of blood; such treatment is useful in Eale's disease, diabetic retinopathy, and hemangiomata. The originally used ruby laser is now generally replaced by the cw argon-ion laser, whose blue and green light is better absorbed by the red tissue.

Wavelength-tunable dye lasers may offer advantages in some of these applications. Their output can be tailored to interact preferentially with specific tissues, in particular when aided by selectively staining dyes.

5.9. Holography

Wavelength-tunable dye lasers offer promising new possibilities in optical holography. It has already been demonstrated that multiline dye lasers can greatly improve the resolution of holographic contour mapping and permit the generation of contour lines with arbitrary separation.

Holography, or the photographic recording and reconstruction of optical wavefronts by means of a coherent reference wave, was first demonstrated by Gabor in 1948. Since the advent of strong, coherent laser sources as a prerequisite for practical applications of this technique, holography has become one of the most actively pursued fields in optical research and development. Two compilations of together almost 2300 articles, books, patents and periodicals relating to holography through 1972 have been published (KALLARD, 1968, 1972). Some of the most recent monographs include books by SMITH (1969), CAULFIELD and LU (1970) and LEHMANN (1970). In 1972, Gabor was awarded the Nobel Prize in physics for his pioneering invention.

The basic principles of holography are simple: the amplitude and phase distribution of a monochromatic optical wavefront falling onto a photographic plate are recorded by superimposing a coherent reference wave, thus forming a complex optical interference pattern. If the processed plate or hologram is later illuminated by coherent light corresponding to the reference wave, the original wavefront with all its optical information is reconstructed by diffraction from the recorded interference pattern. Highly realistic „three-dimensional" photographs can be made by holography, even without imaging lenses. Other areas of application, so far mostly restricted to the laboratory, include interferometry, testing of optical elements, microscopy with potentially much improved depth of view, correction of lens aberrations, information storage, projection printing and micro-imaging, and optical data processing by spatial filtering, e.g. for pattern recognition or image deblurring. High-power dye lasers with narrow bandwidth, clean mode structure, and submicrosecond pulse length could resolve many of the stability problems experienced during exposure, in particular with large-scale holography, and their freely selectable wavelength could be of advantage in many applications of holography.

5.9.1. Holographic Contour Mapping

HILDEBRANDT and HAINES (1967) proposed two different methods for the holographic contour mapping of three-dimensional objects. The first, or two-source method requires only one laser wavelength: two exposures of one object are superimposed on the same hologram, with slight variation in the position of the illuminating source. The reconstruction yields two images, identical in size and position, but with differences in phase. The interference between these two images makes the object appear as if covered with bright and dark bands, the fringes of two interfering point sources. If the object is illuminated with plane waves and viewed at right angles to the direction of illumination, the interference fringes appear as eqidistant contour lines of equal depth. An obvious disadvantage is the possible occurrence of large shadowed areas without contour lines. This drawback is avoided in the second or two-wavelength method. Here the hologram of an object is recorded simultaneously or consecutively with two different laser wavelengths, λ_1 and λ_2. In the reconstruction with one wavelength, two images appear with slightly different size and position. The resulting interference bands correspond to cross sections of the object with ellipsoids or, in limiting cases, with spherical or planar surfaces perpendicular to the direction of illumination. If the illumination source is placed anywhere along the desired line of sight, then the range separation between fringes along this line of sight becomes $\Delta h = \lambda_1 \lambda_2 / 2 (\lambda_1 - \lambda_2)$. The area over which fringes can be observed is restricted by the difference in image size but mapping of large objects is made possible by imaging the object into or close to the holographic plate (ZELENKA and VARNER, 1968).

SCHMIDT and coworkers (1971, 1973) have demonstrated that dye lasers offer two decisive advantages for this method of contour mapping:

1) The depth difference between the contour fringes can be varied by adjusting the wavelength difference between the dye-laser lines;

2) The use of more than two laser wavelengths in equidistant spacing gives narrow multiple-beam interference fringes instead ot two-beam fringes with cosine intensity distribution. Such multiple-line operation is easily accomplished with a Fabry-Perot interferometer with periodic transmission maxima placed inside the dye laser cavity. The interferometer spacing is exactly equal to the depth difference of successive contour lines.

In initial experiments, SCHMIDT and FERCHER(1971) used a flashlamp-pumped dye laser for the contour mapping of small objects (coins). An intracavity Fabry-Perot interferometer restricted laser oscillation to 5 lines of $1.4 \, \text{cm}^{-1}$ separation with a linewidth of less than $0.02 \, \text{cm}^{-1}$. About 1 kW peak power was generated in the TEM_{00} mode. The coin was imaged by a 1 : 1 telescope with angle-limiting aperture onto the photographic plate, and the reference wave was directed at an angle through a similar telescope. Holograms with well-defined contour fringes were recorded over a depth range of 10 mm, limited by the coherence length of the dye laser light.

In later experiments, SCHMIDT et al. (1973) used an argon-laser-pumped cw dye laser with 10—100 mW power in the TEM_{00} mode. Fabry-Perot etalons of different thickness provided multiline operation, and the spectrum could be restricted with an additional interference filter. Approximately true depth contours of a small toy propeller were recorded with a setup as shown in Fig. 5.20. The path lengths for illuminating light and reference light were approximately equalized with the help of two additional mirrors. Figure 5.21 shows contour lines with

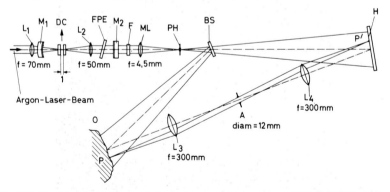

Fig. 5.20. Experimental setup for holographic contour mapping of small objects with a multiline cw-dye laser. The spacing of the contour lines is determined by the tilted Fabry Perot interferometer FPE inside the dye laser cavity, formed by mirrors M_1 and M_2. After spatial filtering (lens ML and pinhole PH), part of the dye laser light is directed by the beamsplitter BS into the object O, and part illuminates the photographic plate P as a reference wave. The lenses L_3 and L_4 with angle-limiting aperture A form an image of the object near the photographic plate. (SCHMIDT et al., 1973)

a

b

Fig. 5.21. Holographic contour lines on a toy propeller. The depth distances between successive lines are 1.5 mm (a) and 0.37 mm (b). (SCHMIDT et al., 1973)

1.5 and 0.37 mm depth difference, generated with solid etalons of 1 mm or 0.25 mm thickness. The laser spectrum contained five lines with a spacing of 0.2 or 0.4 nm. The object size in this scheme is limited by the lens diameters. Contour maps of larger objects could be produced by projecting a ten times reduced image of the object onto the holographic plate. This method yields contours of constant range rather than of constant depth. This is not a major disadvantage, since many applications require only relative measurements, i.e. the comparison of the surface shape with a „master". It is also possible to obtain the true shape by computer evaluation of the fringe pattern. Very sharp contour lines of 6 mm depth separation could be obtained when the dye laser oscillated on 17 lines. A hologram of a Kaplan turbine blade of 40 cm extension required only 10 sec exposure time at 10 mW laser power when the object was coated with a film which scattered light back into the direction of incidence.

Numerous practical applications can be envisioned for holographic contour mapping with dye lasers. Contour maps of the fundus of the eye, for example, would be valuable for medical diagnosis. Pulsed dye lasers of high output power and good coherence length should permit large-scale contour holography of objects like automobiles or sculptures.

5.10. Communications, Integrated Optics

Serveral researches have successfully operated thin-film dye-laser oscillators and related devices, utilizing the waveguiding properties of thin dielectric films. Such devices hold great promise as active elements in integrated optical networks on which future communications systems may be based, though a rather serious difficulty may be the photobleaching of organic dye molecules, which limits the life of such active thin-film waveguides.

The potential of lasers for communications and data transmission was early recognized and strongly emphasized, though much research and development remains to be done before laser systems can be competitive with the older, well-established techniques. A very promising recent advance is the development of waveguiding cladded glass fibers with losses of only 10 dB/km or less. With the help of such fibers it is possible to transmit modulated laser light over long distances, without atmospheric disturbances, and to utilize the huge carrier capacity of electromagnetic waves at optical frequencies to transmit TV programs, telephone conversations, or computer data.

The advent of lasers has initiated immense advances in the field of optoelectronics, in particular in the development of light modulators, frequency mixers, and other nonlinear devices which can be useful for optical communications. But it is only since 1968 that an increasing number of workers have been studying the use of thin-film technology and integrated circuit techniques in the construction of small, waveguiding linear and nonlinear optical elements and their integration into compact, rugged and economical optical systems. The development of this newly emerged and vigorously growing field of integrated optics, the underlying optical principles and the present technological state of the art are reviewed in a very clear paper bei TIEN(1971) and in a report on the first topical meeting on integrated optics and guided waves, held in early 1972 (POLE et al., 1972).

Very high light intensities can be maintained with moderate power in optical waveguides of small cross section. This is of particular interest for the development of nonlinear optical devices, such as modulators, switches, frequency mixers etc. Practical progress has been limited in this area by the difficulty of growing thin crystalline films of good optical quality.

Thin-film dye laser amplifiers and oscillators are especially attractive as elements in integrated optical systems because they can eliminate the need for elements, such as prism or grating couplers, to couple light from the outside into waveguiding films. Moreover, their nonlinear saturation behavior may enable them to take over some of the tasks of other nonlinear devices.

5.10.1. Active Waveguides: Thin-Film Dye Lasers

It is easy to prepare a dye laser structure having transverse dimensions of a few micrometers which supports only a small number of low-order waveguide modes. Modes of low losses are known in straight dielectric waveguides, even if the embedding medium has a higher index of refraction than the guiding core, as in the recently demonstrated hollow waveguide gas lasers. BURLAMACCHI and PRATESI(1973) have utilized this phenomenon in a flashlamp-pumped superradiant dye laser, contained in a small-bore glass capillary. In general, it is more desirable to use a surrounding dielectric medium of lower refractive index to permit waveguiding by total internal reflection, and the wide choice of available liquid and solid dye-laser host materials makes it easy to meet this condition. Such waveguiding liquid dye lasers have been constructed by using benzyl alcohol ($n = 1.538$) as a solvent and filling the liquid dye solution into thin glass capillaries (IPPEN et al., 1971; ZEIDLER, 1971) or sandwiching a liquid dye film between flat glass substrates (ZEIDLER, 1971).

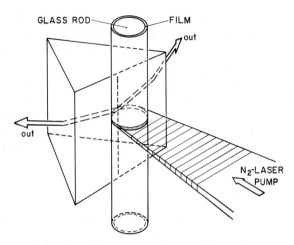

Fig. 5.22. Waveguiding thin-film ring laser, pumped by a nitrogen laser. The active medium is a 0.8 µm thick polyurethane film, doped with rhodamine 6G, coated on the outside of a glass rod. (WEBER and ULRICH, 1971)

WEBER and ULRICH (1971) have reported the successful operation of a solid thin-film dye laser. They produced the active waveguiding structure by coating glass substrates with a thin film of polyurethane ($n = 1.55$) doped with $8 \cdot 10^{-3}$ M/l rhodamine 6G. At a typical thickness of 0.8 µm, such a film can support only the fundamental TE_0 and TM_0 modes. A gain as high as 100 dB/cm was obtained when the film was pumped by a pulsed N_2 laser. In order to provide feedback for laser oscillation, the doped light-guiding film was applied on the surface of a cylindrical glass rod of 5 mm diameter, as shown in Fig. 5.22. In this way a closed optical path is established along any circumference of the rod. A narrow circumferential strip of the beam is illuminated by a sheetlike beam of the N_2 laser. The dye-laser light is coupled out via its evanescent wave by a closely spaced prism. Two beams are obtained, corresponding to the two opposite directions of rotation. A peak power of 100 W was measured in each beam when the film absorbed about 1 kW of the incident pump light. The laser operated near 620 nm with a bandwidth of 11 nm. Individual axial modes of the ring resonator could be resolved with a Fabry-Perot interferometer, confirming the feedback around the rod.

An entirely different approach to waveguiding dye lasers is realized in the evanescent-field-pumped dye laser, as reported by IPPEN and SHANK(1972). Here the dye molecules are not incorporated in the waveguide, but are located in the surrounding (liquid) medium. They are pumped by the evanescent wave of the pump-laser light travelling through the waveguide, and they radiate by stimulated emission into a waveguide mode. In the experiment the pumplight of a frequency-doubled Nd : glass laser was coupled via a prism into a thin waveguiding glass film, covered by a solution of rhodamine 6G in a benzyl alcohol glycerol mixture. The superradiant dye laser output was measured for different liquid refractive indices and different pump powers. The reported scheme appears particularly attractive because it is easy to replenish photo-bleached dye molecules. A theoretical estimate indicates that it should be possible to construct a rhodamine 6 G dye laser with a waveguide cross section of $1 \times 3 \mu m^2$, which would exhibit a threshold of only 5 mW at a round-trip loss of 10%.

5.10.2. Dye Lasers with Distributed Feedback

A very versatile way to obtain optical feedback in waveguiding structures is realized in dye-laser oscillators with distributed feedback, as demonstrated by BJORKHOLM and SHANK(1972). An optical parameter of the active film, such as its refractive index, gain, or thickness, is spatially modulated in a grating-like fashion, and highly wavelength-selective feedback is obtained by Bragg backscattering from this periodic structure.

Dye-laser oscillators with feedback distributed throughout the (nonwaveguiding) active medium were first reported by KOGELNICK and SHANK(1971), who used a small holographic phase grating, produced in a dichromated gelatin film on a glass substrate, and doped it with rhodamine 6G by soaking it in a dye solution. When side-pumped by a N_2 laser, this compact distributed-feedback laser $(10 \times 0.1$ mm$^2)$ oscillated in a band of less than 0.05 nm width near 630 nm. KAMINOV et al. (1971) described a longitudinally pumped monochromatic dye laser which consists of a pair of Bragg phase gratings, photodielectrically induced inside a bulk sample of polymethylmethacrylate (PMMA) doped with rhodamine 6G, as shown in Fig. 1.50. The gratings produced by the permanent change in the refractive index of PMMA induced by irradiation with interfering UV laser beams had dimensions of $2 \times 2 \times 2$ mm^3 and a maximum reflectivity of 20%. The pump laser was a frequency-doubled Nd : YAG laser with 1.3 MW peak power at 530 nm. A dye-laser output of 65 kW in three axial modes, covering less than 0.01 nm, was obtained. The wavelength is not temperature-sensitive because thermally induced changes of refractive index are compensated for by changes in the grating constant. Stress tuning over a range up to 10 nm appears possible, however.

SHANK et al. (1971) obtained broadly tunable dye-laser action with 0.03—0.001 nm linewidth and 36 kW peak power from a mirrorless dye laser system with distributed feedback. The laser gain and refractive

index is spatially modulated simply by side-pumping the 1.6 cm-long dye
cell with the fringes formed by the interference of two coherent light
beams (see Fig. 1.47). The pump laser is a frequency-doubled Q-switched
ruby laser of 180 kW peak power. With rhodamine 6G the wavelength
could be tuned over a range of 64 nm by changing either the pump angle
or the refractive index of the solvent. Narrowband laser oscillation from
such a system was also observed when the period of the susceptibility
modulation was two or three times larger than the oscillation
wavelength (BJORKHOLM and SHANK, 1972). This higher-order operation
is explained by the very high gain of laser-pumped dyes and by possible
harmonic frequencies in the spatial gain modulation due to saturation.
Higher-order Bragg scattering from a thick, sinusoidally modulated
grating is very weak compared with first-order scattering.

FORK and KAPLAN (1972) reported on a distributed-feedback dye laser
with a variable phase grating which can be optically written into photodimer optical memory material with the interference fringes of a 364 nm
Ar$^+$ laser, or erased with the 313 nm light of a Hg arc lamp. The laser
material is a solid PMMA host, doped with rhodamine 6G and photodimers of acridizinium ethylhexanesulfonate. The photodimer can be
broken with the erasing light and remade with the writing laser light of
longer wavelength. Writing times as short as 3 nsec appear possible with
high-power pulsed lasers. A gain of 8 cm^{-1}, twice above threshold, was
obtained, when the small active volume (1 cm $\times$ 80 μm $\times$ 10 μm) was
pumped with a pulsed neon laser at 540 nm with 10 kW peak power. The
narrowband dye-laser output consists of several peaks, some less than
1.2 GHz wide.

KOGELNIK and SHANK (1972) have given a theoretical analysis of a
laser with distributed feedback in bulk material. In this model, two
counterrunning waves are coupled by backward Bragg scattering from a
small spatial modulation of refractive index and gain. Saturation effects
are neglected, and numerical results are derived for mode spectra,
threshold gain, mode patterns and spectral selectivity.

Dye-laser oscillators with distributed feedback in thin-film optical
waveguides have recently been investigated by a number of workers.
BJORKHOLM and SHANK (1972) have produced active waveguides by
coating flat glass substrates with polyurethane film, doped with rhodamine 6G. Distributed feedback was generated, as in the earlier experiments with liquids, by pumping with spatially modulated light provided
by a frequency-doubled ruby laser. The output of the waveguide laser
was observed by guiding the light scattered from the end of the waveguide through a lens into a spectrograph. Two laser lines, separated by
about 1 nm, are emitted by thin single-mode waveguides; they correspond
to the two fundamental modes TE_0 and TM. The linewidth of 0.05 to

0.5 nm is noticeable larger than that of the previous liquid lasers, due to film imperfections. Thicker films which can support many modes with different phase velocities exhibit a more complex and somewhat un-predictable spectrum. Comparison with a theoretical model indicates that the behavior of thin-film waveguide lasers is in general well under-stood. BJORKHOLM and SHANK suggested that the study of active wave-guides can serve as a convenient and quick method to measure guide parameters, such as film thickness, thickness variations, or refractive index, without mechanical contact and physical distortions.

ZORY (1972) has operated successfully a distributed feedback thin-film laser, consisting of a film of liquid organic dye solution, held between a blazed reflection grating and an optical flat. Feedback is provided by the fourth spatial harmonic of the grating, and the output is coupled out from the grating under the first-order angle in the form of a leaky wave.

SCHINKE et al. (1972) fabricated a somewhat similar thin-film dye laser with distributed feedback, produced by a periodic perturbation of the guide height. Using photoresist and ion-milling techniques, these authors produced gratings with a period of about 0.2 μm and a peak height of about 0.1 μm on fused silica substrates. A film of polyurethane, doped with rhodamine 6 G and pumped by a N_2 laser, formed the active waveguide. The laser output, extracted with a prism coupler, has a line-width of less than 0.1 nm. Oscillation thresholds, measured for several different lasers, are in agreement with calculated feedback parameters. The milled quartz gratings are not only durable, they also can provide a substantial feedback, sufficient to reach laser threshold with low-gain materials. At a typical feedback parameter of 300 cm^{-1}, a gain of only 10^{-4} cm^{-1} would be required for dye-laser oscillation in a 1 cm-long structure without absorption losses.

Distributed feedback in a waveguiding thin-film laser can also be produced by periodic variations of refractive index in an adjacent dielectric medium of lower refractive index. The feedback mechanism then involves the evanescent waves of the waveguide. The feasibility of such a scheme was demonstrated by HILL and WATANABE (1972), who sandwiched a liquid-dye-solution film between a holographic phase grating, produced in dichromated gelatin, and an optical flat. The active medium had a thickness of 15 to 100 μm and consisted of a solution of rhodamine 6 G in a mixture of benzyl alcohol and ethanol. The dye laser was side-pumped by a N_2 laser. It was also possible to achieve laser oscillation in different modes of operation. In particular, if the dye solu-tion had a lower index of refraction than the gelatin grating, the gelatin film acted as waveguide with distributed feedback and gain was provided by the bordering dye medium via the evanescent waves.

The reported experiments certainly demonstrate the viability of the concept of waveguide dye lasers. The availability of compact, rugged inexpensive narrowband dye-laser oscillators with integrated resonator structure would certainly be of interest for a variety of applications beyond the field of integrated optics, but the practical usefulness of such dye lasers will remain very limited unless ways can be found to reduce the photobleaching of the organic dye molecules, which severely limits the life of the present devices. IPPEN et al. (1971) have measured bleaching cross sections for several dyes in liquid solution, and KAMINOV et al. (1972) reported equivalent measurements for dye molecules in solid hosts. The quantitative results are very similar in both cases. Rhodamine 6G is the most stable of the dyes tested. One rhodamine molecule can emit on average about $2 \cdot 10^6$ laser photons, before it is destroyed by some unknown bleaching process, perhaps due to a chemical reaction in its excited state. Impurities in the host material can much reduce its useful life. The bleaching rate is a linear function of the pump intensity up to power densities of $10^7 \, \mathrm{W/cm^2}$, which rules out any significant bleaching by two-photon processes. At a typical threshold pump intensity of $10^5 \, \mathrm{W/cm^2}$, a rhodamine 6G molecule has a life of only about 40 msec, and a pulsed laser will operate only for some 10^3 to 10^4 shots, unless the dye molecules are replenished.

One possible, though not very convenient, solution of this problem may be to cool the dye laser material, as indicated by experiments of FORK and KAPLAN (1972). Rhodamine 6G molecules in a PMMA matrix, pumped in the green by a frequency-doubled Nd : YAG laser, can emit more than 10^7 laser photons without bleaching, provided the host is cooled below 200° K. This increased resistance to photodegratation may be due to a reduced mobility of impurities or a more rigid polymer cage around the active molecule. Only some 20% of the dye molecules seem to be subject to bleaching at low temperatures, so making it unlikely that a thermal excitation of the dye molecules is responsible for the photochemical destruction.

5.11. Light Amplification

For most laser applications the amplifying medium is enclosed in an optical cavity for feedback and mode selection, i.e. it is part of a laser oscillator. It is possible, however, to use the amplifying medium without resonator simply as a traveling-wave light amplifier. The wide bandwidth and potentially very large gain of dye-laser amplifiers makes them particulary attractive for such apllications. For high dye concentrations on the order of $10^{-2}\,M/l$ and strong excitation, where 10% of the molecules are pumped into the excited singlet state, one expects theoretically gain coefficients as large as 300 dB/cm. Measurements by SHANK et al. (1970) and HÄNSCH et al. (1971) have confirmed that the gain coefficients of dye solutions pumped by a N_2 laser can be on the order of 20—230 dB/cm. The gain of flashlamp-pumped dye lasers is generally considerably lower (HUTH, 1970).

The usefulness of laser amplifiers is somewhat limited by the background of amplified spontaneous emission. From basic principles of quantum physics it can be derived that the noise in the output of a laser amplifier of high gain corresponds to an input of at least one photon per mode, or to an input of blackbody radiation of more than 20000° K (KOGELNIK and YARIV, 1964). Hence light from ordinary thermal sources, even direct sunlight, added to the input of a laser amplifier, would give a negligible increase of the output above the noise level. Only light from coherent sources will produce useful signals at the output.

The maximum useful single-pass gain is also determined by the amplified spontaneous emission, because a high noise level at the output reduces the gain by saturation. Typical saturation parameters are on the order of $100\,kW/cm^2$ to $1\,MW/cm^2$. The intensity of the amplified spontaneous emission must be kept well below this saturation intensity if „superradiant" emission is to be avoided. Amplification factors of up to 1000 have been realized in practice without violating this condition. Much higher gains should be possible if the dye-laser amplifier is subdivided into several stages and the number of modes of the transmitted noise radiation reduced by spatially limiting apertures and spectral filters.

Unlike solid-state lasers, dye lasers cannot store pump energy for longer than a few nanoseconds, i.e. the lifetime of the excited singlet state. Dye-laser amplifiers are consequently not very suitable for the generation of extremely high pulse powers, but they are satisfactory for amplifying weak signals to moderately high intensities.

Experimental studies of dye-laser amplifiers have been reported by many workers. The first measurements of broadband light amplification in organic dyes were made by BASS and DEUTSCH(1967). They used a Q-switched ruby laser to pump the dye amplifier, operating in the range 700 nm to 850 nm, and used the Raman-shifted ruby output to test the amplifying characteristics of the dyes. The first gain measurement of a flashlamp-pumped rhodamine 6G dye laser was reported by HUTH(1970), yielding a gain coefficient of 95 dB/m. SHANK et al.(1970b) measured the gain of several N_2 laser-pumped dyes by comparing the amplified spontaneous emission for different length of the amplifying medium, according to a technique of Silfvast and Deech. Transient gain measurements of flashlamp-pumped rhodamine 6G solutions were made by STROME (1972).

5.11.1. Dye-Laser Oscillator-Amplifier Systems

Laser- and flashlamp-pumped dye-laser amplifiers have been used to boost the power of a weak but clean laser oscillator and to improve the spectral brightness and spatial radiance of the laser output.

A theoretical model of a flashlamp-pumped high-gain dye laser amplifier has been studied by GASSMAN and WEBER(1971). These authors solve a system of coupled rate equations, taking triplet-triplet state absorption into account and calculate the maximum gain in the limit of weak pumping as a function of the amplifier geometry. FLAMANT and MEYER(1973) analyzed a similar model in the limit of high pump intensities and substantial saturation where the laser gain no longer depends exponentially on the amplifier length.

N_2-laser-pumped dye-laser oscillator–amplifier systems with narrow bandwidth and high output power have been described by HÄNSCH and SCHAWLOW(1970) and by ITZKAN and CUNNINGHAM(1972). The latter authors use two flow dye cells, pumped by two synchronized N_2 lasers. The cavity of the low-power oscillator consists of a grating in Littrow configuration and a semitransparent mirror, with an additional tilted Fabry-Perot interferometer inserted. An aperture between oscillator and amplifier reduces the beam divergence to about three times the diffraction-limited value. With this system, peak powers of 5—50 kW at 5 nsec pulse width and 2 pm bandwidth could be generated throughout the visible spectrum from 360—670 nm with overall efficiences reaching 25%.

FLAMANT and MEYER(1971) have constructed a flashlamp-pumped dye-laser oscillator with multistage dye-laser amplifier. Each of the six amplifier stages consists of an 8 cm-long cylindrical dye cell, pumped by a linear flashlamp in an elliptical reflector. A gain of 700 was measured for rhodamine 6G solution near 590 nm, and an energy gain of 6 was reported for an output of 60 mJ. With a later version of a flashlamp-pumped dye-laser amplifier FLAMANT and MEYER (1973) achieved a power amplification of 230 and outputs up to 3.7 MW/cm^2 or 1 joule. CARLSTEN and McILRATH (1973) described an oscillator-amplifier dye laser, pumped by a single Q-switched ruby laser, for applications in the spectroscopy of excited atomic states. The system generates pulses of 5 MW peak power and 30 nsec width with a linewidth of 50 pm. A gain of 20 and a conversion efficiency of 12% were obtained near 790 nm.

5.11.2. Intensity Stabilization by Saturated Amplification

CURRY et al. (1973) have proposed a novel scheme for stabilizing the intensity of a fluctuating laser source, using a partly saturated high-gain traveling-wave dye-laser amplifier. Such a stabilization scheme is particularly desirable when the output of a short-pulse dye laser is sent through an ultranarrow external passband filter, in order to obtain a bandwidth of the order of a few MHz, as required for high-resolution saturation spectroscopy of gaseous absorption lines. The spectral filtering reveals large spectral fluctuations within the laser bandwidth, as expected for highly amplified spontaneous emission with essentially Gaussian statistics. The relative intensity fluctuations $F = \Delta I/I$ can obviously be reduced by a subsequent dye-laser amplifier whose gain for the stronger pulses is diminished by saturation.

CURRY and coworkers (1973) have made a theoretical analysis of the merits of such a stabilizing amplifier, using a rate equation model in the steady-state approximation. The stabilization factor $S = d\ln(I_{in})/d\ln(I_{out})$ has an upper limit corresponding to the natural logarithm of the total gain G, so that even a very optimistic amplifier gain of 100 dB one would hardly give a stabilization factor higher than $S = 20$. The reason for this poor performance is easy to understand: There is only a small section of the amplifier where the intensity of the laser source is near the value I_{sat} from which one expects the best stabilization properties. In the first section of the amplifier the intensity is generally too weak to saturate and in the final section it has increased to a level where in effect it „bleaches" the medium and passes through without further amplification or stabilization.

The stabilization factor can be increased by several orders of magnitude to a maximum value $S_{max} = G^{1/2}$ by adding to the amplifying medium a non-saturable absorber whose absorption coefficient equals half the unsaturated amplifier gain coefficients. The medium then exhibits gain for smaller light intensities $(I < I_{sat})$ and absorption for larges intensities $(I > I_{sat})$ and in any event the intensity will tend to approach the limiting value I_{sat} at which the medium operates in a region of maximum stabilization. Instead of using a continuous absorber, it is possible to subdivide the amplifier into separate stages and to insert attenuating filters between them. CURRY et al. reported an experiment in which a two-stage dye-laser amplifier with inserted attenuator, pumped by a nitrogen laser, is used to amplify and stabilize the intensity of an ultra-narrowband pulsed dye laser (see Fig. 5.23). Each amplifier stage, operating with rhodamine 6G near 600 nm, had an unsaturated single-pass gain of about 300. The output peak power was on the order of 1 kW and changed by a factor of less than four when the input was varied over

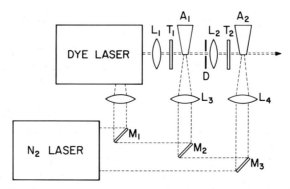

Fig. 5.23. Two stage dye laser amplifier for intensity stabilisation of a fluctuating dye laser. The dye laser oscillator and the two amplifier cells A_1 and A_2 are side-pumped by the same nitrogen laser via mirrors M and quartz lenses L. The amplifier input intensity can be varied with the attenuator T_1. The stabilization factor is optimized with a second attenuator T_2 inserted between the amplifier stages. (CURRY et al., 1973)

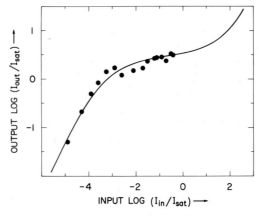

Fig. 5.24. Output of two-stage dye-laser amplifier with inserted attenuator versus input. The output changes by a factor less than of four while the input intensity changes over three orders of magnitude. (CURRY et al., 1973)

three orders of magnitude (Fig. 5.24). However, the amplifier broadened the linewidth of the incident laser radiation from 10^{-5} to 10^{-4} nm, as expected from its short pulse width of only 7 nsec.

5.11.3. Dye-Laser Amplifiers with Feedback

Dye-laser amplifiers with optical feedback by a cavity can offer certain advantages over traveling-wave laser amplifiers as boosters for weak laser oscillators, so long as accurate preservation of the light frequency is not required. In particular, it is possible to obtain a much higher effective power gain from a given laser amplifier in this way. It may be a disadvantage that it is generally rather difficult to keep such a device below the oscillation threshold; in the absence of a sufficiently strong input signal the amplifier will operate as a broadband laser oscillator.

The practical feasibility of the optical feedback scheme was demonstrated by ERICKSON and SZABO(1971) who injected the 514.5 nm light of a 50 mW argon ion laser into the cavity of a broadband 4-methyl-umbelliferone dye laser, side-pumped by a N_2 laser. With monochromatic injection, the dye-laser output is spectrally narrowed from 40 nm into one to three axial cavity modes of 1.6 pm linewidth and 7 GHz spacing (Fig. 5.25). The saturated peak power was about 4 kW. VREHEN and BREIMER(1972) obtained similar results by injecting the light of a 15 mW He–Ne laser, operating at 632.8 nm, into the cavity of a longitudinally pumped cresyl violet dye laser. The authors explain by means of a simplified theoretical model why the frequency of the injected radiation need not exactly coincide with any of the cavity mode frequencies, provided the inital gain is sufficiently large. MAGYAR and SCHNEIDER-MUNTAU(1972) achieved an energy output of 600 mJ in a bandwidth of 10 pm and an effective power gain of 4000 by sending the light of a small, flashlamp-pumped coaxial rhodamine 6G dye laser into a large, close-coupled flashlamp-pumped dye-laser oscillator.

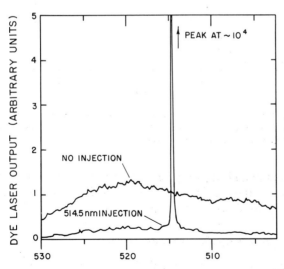

Fig. 5.25. Spectrum of a nitrogen-laser-pumped coumarin dye laser without and with injection of monochromatic 514.5nm light from an argon-ion laser. (ERICKSON and SZABO, 1971)

5.11.4. Image Amplification

HÄNSCH et al. (1971) have demonstrated a different application of light amplification with dye lasers: they constructed a compact, pulsed dye-laser image amplifier with a single-pass gain up to 30 dB and with diffraction-limited optical resolution. Various attempts had been made previously to use laser amplifiers to transmit and intensify optical images. Experimental results obtained with traveling-wave ruby-laser amplifiers have been reported, but gain and resolution were moderate. Numerous practical applications can be envisioned for the above dye-laser device. Because of unavoidable quantum noise, laser image amplifiers obviously cannot compete with conventional image-intensifying systems when it comes to amplifying low light intensities from thermal sources. On the other hand, they offer the unique advantage of preserving within certain limits not only amplitude and direction but also color (frequency) and phase of the incoming light. In addition, high output powers can be obtained, and a time resolution down to subpicoseconds appears feasible.

The prototype dye-laser image amplifier is rather simple device in construction. A photograph of part of the experimental setup is shown in Fig. 5.26. Two dye cells are used, one acting as the amplifier and the other as an oscillator or light source. The cell windows are sealed on at a wedge angle to prevent feedback and the formation of optical cavities. Both cells are filled with the same dye solution, e.g. rhodamine 6G in

Fig. 5.26. Prototype of dye-laser image amplifier. (HÄNSCH et al., 1971)

ethanol, and are pumped simultaneously by the same repetitively pulsed N_2 laser. The ultraviolet output of the laser pump source is focused by spherical quartz lenses from the side into a thin line at the inner wall of the cells, which have an active length of 1.3 mm. Part of the dim fluorescent light of the oscillator cell which shows a noticeable preference for near-axial propagation is collected by a field lens and focused by an additional multielement photographic lens ($f = 55$ mm) into the active volume of the amplifier cell. After being amplified in a single pass, the light emerges in the form of a bright cone at the other end, illuminating a circular area on a projection screen. Any object, such as a wire mesh or a transparency, held in the object plane of the photographic lens throws a bright projected image on the screen, despite the faint illumination of the object.

Saturation reduces the amplifier gain at higher output powers. For rhodamine 6G, the gain dropped from 23 dB/mm to 14 dB/mm when the total pulse output reached $25\,\mu J$ or some 25% of the pump energy. Despite this gain saturation, the image contrast is not much changed in a single pass, because in essence the amplifier processes the Fourier transform of the optical image, and the active volume is shared by light from different object points.

A spatial resolution of 200 line pairs per 0.4 rad could be obtained with careful optical adjustment. This resolution appears to be limited only by diffraction since the small cross section of the optically excited dye region acts as an active pinhole. The short pulse width of a few nanoseconds prevents blurring of the image by thermally induced changes in refractive index associated with a relatively slow thermal volume expansion after absorption of the pump light.

The ability to read optical information at relatively low intensities and to use it at a much higher intensity level has a rather general appeal and is of special advantage where an object might be damaged by intense illumination but where the light must be bright enough for the object to be seen or photographed. Thus dye-laser image amplifiers could find applications in microscopy or microfilm projection, or in (color) movie projection with potentially greatly reduced frame size. Full color operation could, for example, be realized by using three dye laser amplifier cells with three different dye solutions in parallel. One can envision other promising uses in numerous fields, including spectroscopy, short-time photography or schlieren optics. Dye-laser image amplifiers could also be useful in electron-beam-controlled display systems, for instance for TV projection, or in large-scale optical information processing, using read-write optical computer memories.

References

ABAKUMOV, G. A., SIMONOV, A. P., FADEEV, V. V., KHARITONOV, L. A., KHOKHLOV, R. V.: (1969a), JETP Letters **9**, 9; Russ.: **9**, 15.

ABAKUMOV, G. A., SIMONOV, A. P., FADEEV, V. V., KASYMDGANOV, M. A., KHARITONOV, L. A., KHOKHLOV, R. V.: (1969b), Opto-Electronics **1**, 205.

AGOSTINI, P., BENSOUSSAN, P., BOULASSIER, J. C.: (1972), VII. International Quantum Electronics Conference, pp. 26: Montreal: Technical Digest.

AKHMANOV, A. G., AKHMANOV, S. A., ZHDANOV, B. V., KOVRIGIN, A. I., PODSOTSKAYA, N. K., KHOKLOV, R. V.: (1969), JETP Letters **10**, 154.

ALEKSEEV, V. A., ANTONOV, I. V., KOROBOV, V. E., MIKHNOV, S. A., PROKUDIN, V. S., SKORTSOV, B. V.: (1972), Soviet J. Quant. Electr. **1**, 643.

AMBARTSUMYAN, R. V., LETOKHOV, V. S.: (1972), Appl. Opt. **11**, 354.

ARISTOV, A. V., VIKTOROVA, E. N., KOZLOVSKII, D. A., KUZIN, V. A.: (1970), Opt. Spectrosc. **28**, 293; Russ.: **28**, 546.

ARISTOV, A. V., KUZIN, V. A.: (1971), Opt. Spectrosc. **30**, 77; Russ.: **30**, 148.

ARMSTRONG, J. A.: (1967), Appl. Phys. Letters **10**, 16.

ARTHURS, E. G., BRADLEY, D. J., RODDIE, A. G.: (1971), Appl. Phys. Letters **19**, 480.

ARTHURS, E. G., BRADLEY, D. J., LIDDY, B., O'NEILL, F., RODDIE, A. G., SIBBETT, W., SLEAT, W. E.: (1972a), Proc. International Conference on High Speed Photography. Nice, France.

ARTHURS, E. G., BRADLEY, D. J., RODDIE, A. G.: (1972b), Appl. Phys. Letters **20**, 125.

ARVAN, KH. L., ZAITSEVA, N. E.: (1961), Opt. Spectrosc. **10**, 137; Russ.: **10**, 272.

ASHKIN, A.: (1970), Phys. Rev. Letters **25**, 1321.

AUSSENEGG, F., SCHUBERT, J.: (1969), Phys. Letters **30**A, 488.

AUSTON, D. H.: (1971), Appl. Phys. Letters **18**, 249.

BAARDSEN, E. L., TERHUNE, R. W.: (1972), Appl. Phys. Letters **21**, 209.

BALTAKOV, F. N., BARIKHIN, B. A., KORNILOV, V. G., MIKHNOV, S. A., RUBINOV, A. N., SUKHANOV, L. V.: (1973), Sov. Phys. Tech. Phys. **17**, 1161.

BARANOVA, E. G.: (1965), Opt. Spectrosc. **18**, 230; Russ.: **18**, 407.

BARGER, R. L., SOREM, M. S., HALL, J. L.: (1973), Frequency Stabilization of a CW Dye Laser, Appl. Phys. Lett. **22**, 573.

BASS, M., DEUTSCH, T. F.: (1967), Appl. Phys. Letters **11**, 89.

BASS, M., DEUTSCH, T. F., WEBER, M. J.: (1968), Appl. Phys. Letters **13**, 120.

BASS, M., STEINFELD, J. I.: (1968), IEEE J. Quant. Electr. **4**, 53.

BATES, B., BRADLEY, D. J., KOHNO, T., YATES, H. W.: (1968), Rev. Sci. Instr. **34**, 476.

BECKER, R. S.: (1969), Theory and Interpretation of Fluorescence and Phosphorescence. New York: Wiley-Interscience.

BEESLY, M.: (1971), Laser and their Applications. New York: Taylor & Francis.

BENEDEK, G. B.: (1969), Polarisation, matière et rayonnement. Paris: Presse Universitaire de France.

BERGMAN, A., DAVID, R., JORTNER, J.: (1972), Optics Commun. **4**, 431.

BEYER, R. A., LINEBERGER, W. C.: (1973), to be published.

BIRKS, J. B.: (1970), Photophysics of Aromatic Molecules. London: Wiley-Interscience.

BIRKS, J. B.: (1972), Chem. Phys. Letters **17**, 370.

BJORKHOLM, J. E., DAMEN, F. C., SHAH, J.: (1971), Optics Commun. **4**, 283.

BJORKHOLM, J. E., SHANK, C. V.: (1972a), IEEE J. Quantum Electr. QE-**8**, 833.

BJORKHOLM, J. E., SHANK, C. V.: (1972b), Appl. Phys. Letters **20**, 306.

BLOEMBERGEN, N.: (1965), Nonlinear Optics. New York: Benjamin.

BLOUNT, E. I., KLAUDER, J. R.: (1969), J. Appl. Phys. **40**, 2874.

BOITEUX, M., DeWITTE, O.: (1970), Appl. Optics **9**, 514.

BOLGER, B., WEYSENFELD, C. H.: (1972), VII. International Quantum Electronics Conference, pp. 11. Montreal: Technical Digest.

BONCH BRUYEVICH, A. M., KHODOVOY, V. A., KHROMOV, V. V.: (1970), JETP Letters **11**, 290.

BONNEAU, R., FAURE, J., JOUSSOT-DUBIEN, J.: (1968), Chem. Phys. Letters **2**, 65.

BOWMAN, M. R., GIBSON, A. J., SANDFORD, M. C.: (1969a), Joint Conference on Lasers and Optoelectronics, Southampton.

BOWMAN, M. R., GIBSON, A. J., SANDFORD, M. C.: (1969b), Nature **221**, 456.

BOYD, G. D., GORDON, J. P.: (1961), Bell Syst. Techn. J. **40**, 489.

BRADLEY, D. J.: (1970), Phys. Bulletin **21**, 116.

BRADLEY, D. J.: (1971), Proc. of the Tech. Programme, Electro-Optics '71, International Conference, Brighton.

BRADLEY, D. J., GALE, G. M., MOORE, M., SMITH, P. D.: (1968a), Phys. Letters **26**A, 378.

BRADLEY, D. J., DURRANT, A. J. F.: (1968), Phys. Letters **27**A, 73.

BRADLEY, D. J., DURRANT, A. J. F., GALE, G. M., MOORE, M., SMITH, P. D.: (1968b), IEEE J. Quant. Electr. QE-**4**, 707.

BRADLEY, D. J., O'NEILL, F.: (1969), J. Opto-Electron. **1**, 69.

BRADLEY, D. J., DURRANT, A. J. F., O'NEILL, F., SUTHERLAND, B.: (1969), Phys. Letters **30**A, 535.

BRADLEY, D. J., EWART, P., NICHOLAS, J. V., SHAW, J. R. D.: (1972), VII. International Quantum Electronics Conference, pp. 58. Montreal: Technical Digest.

BRADLEY, D. J., GALE, G. M., SMITH, P. D.: (1970a), J. Phys. B**3**, 11.

BRADLEY, D. J., GALE, G. M., SMITH, P. D.: (1970b), Nature **225**, 719.

BRADLEY, D. J., LIDDY, B., SLEAT, W. E.: (1971a), Opt. Commun. **2**, 391.

BRADLEY, D. J., LIDDY, B., RODDIE, A. G., SIBBETT, W., SLEAT, W. E.: (1971b), Opt. Commun. **3**, 426.

BRADLEY, D. J., CAUGHEY, W. G. I., VUCUSIC, J. I.: (1971c), Opt. Commun. **4**, 150.

BRADLEY, D. J., NICHOLAS, J. V., SHAW, J. R. D.: (1971d), Appl. Phys. Letters **19**, 172.

BROCK, E. G., CZAVINSKY, P., HORMATS, E., NEDDERMAN, H. C., STRIPE, D., UNTERLEITNER, F.: (1961), J. Chem. Phys. **35**, 759.

BROIDA, H. P., HAYDON, S. C.: (1970), Appl. Phys. Letters **16**, 142.

BROUDE, V. L., MASHKEVICH, V. S., PRIKHOT'KO, A. F., PROKOPYUK, N. F., SOSKIN, M. S.: (1963), Sov. Phys.-Solid State **4**, 2182.

BÜCHER, H., KUHN, H.: (1970), Chem. Phys. Letters **6**, 183.

BUETTNER, A. V., SNAVELY, B. B., PETERSON, O. G.: (1969), in: Molecular Luminescence, p. 403. LIM, E. C. (Ed.). New York: Benjamin.

BURRELL, C. F., KUNZE, H. J.: (1972a), Phys. Rev. Letters **28**, 1.

BURRELL, C. F., KUNZE, H. J.: (1972b), Phys. Rev. Letters **29**, 1445.

BURLAMACCHI, R., PRATESI, R.: (1973), Appl. Phys. Letters **22**, 7.

BYER, R. L.: (1973), Optical parametric oscillators, in: Treatise in Quantum Electronics, RABEN, H., TANG, C. L. (Eds.). New York, London: Academic Press, to be published.

BYER, R. L., HERBST, R. L., KILDAL, H., LEVENSON, M. D.: (1971), Appl. Phys. Letters **20**, 463.

CALVERT, J. G., PITTS, J. N., Jr.: (1966), Photochemistry. New York: Wiley.

CAPELLE, G., SAKURAI, K., BROIDA, H. P.: (1971), J. Chem. Phys. **54**, 1728.

CARLSTEN, J. L., McILLRATH, T. J.: (1973), Opt. Commun., to be published.

CAULFIELD, H. J., LU, S.: (1970). The Applications of Holography. New York: Wiley.

CHANCE, B., ERECINSKA, M.: (1971), Arch. Biochem. and Biophysics **143**, 675.

CHANDROSS, E. A., VISCO, R. E.: (1964), J. Am. Chem. Soc. **86**, 5350.

CHARSCHAN, S. S., ed.: (1972). Lasers in Industry. New York: Van Nostrand Reinhold.

CIRKEL, H. J., RINGWELSKI, L., SCHÄFER, F. P.: (1972), Z. Physik. Chemie Neue Folge **81**, 158.

CLAESSON, S., LINDQUIST, L.: (1958), Arkiv Kemi **12**, 1.

COLOUR INDEX: (1971), Third Edition (The Society of Dyers and Colourists, Bradford, England), Vol. 5.

CROZET, P., KIRKIACHARIAN, B. S., SOULA, C., MEYER, Y. H.: (1971), J. Chim. Phys. **68**, 1388.

CROZET, P., MEYER, Y.: (1970), C. R. Acad. Sci. Paris **271**, 718.

CURRY, S. M., CUBEDDU, R., HÄNSCH, T. W.: (1973), Appl. Phys. **1**, 153.

DAMEN, T. C., SHAH, J.: (1971), Phys. Rev. Letters **27**, 1506.

DANIELMEYER, H. G.: (1971), J. Appl. Phys. **42**, 3125.

DECKER, C. D., TITTEL, F. K.: (1973), Appl. Phys. Letters, to be published.

DEMAS, J. N., CROSBY, G. A.: (1971), J. Phys. Chem. **75**, 991.

DENTON, M. B., MALMSTADT, H. V.: (1971), Appl. Phys. Letters **18**, 485.

DERKACHEVA, L. D., KRYMOVA, A. I., VOMPE, A. F., LEVKOEV, I. I.: (1968a), Opt. Spectrosc. **25**, 404; Rus.: **25**, 723.

DERKACHEVA, L. D., KRYMOVA, A. I., MALYSHEV, V. I., MARKIN, A. S.: (1968b), JETP Letters **7**, 362; Russ.: **7**, 468.

DERKACHEVA, L. D., SOKOLOVSKAYA, A. I.: (1968), Opt. Spectrosc. **25**, 244.

DERKACHYOVA, L. D., KRYMOVA, A. I., MALISHEV, V. I., MARKIN, A. S.: (1969), Opt. Spectrosc. **26**, 572.

DEUTSCH, T. F., BASS, M.: (1969), IEEE J. Quant. Electr. QE-**5**, 260.

DEVLIN, G. E., DAVIS, J. L., CHASE, L., GESCHWIND, S.: (1971), Appl. Phys. Letters **19**, 138.

DEWEY Jr., C. F.: (1971), Excitation of gases using wavelength tunable lasers, in: Modern Optical Methods in Gas Dynamics Research, DOSANJH, S. S. (Ed.). New York: Plenum Press.

DEWEY Jr., C. F., HOCKER, L. O.: (1971), Appl. Phys. Letters **18**, 58.

DIENES, A., SHANK, C. V., TROZZOLO, A. M.: (1970), Appl. Phys. Letters **17**, 189.

DIENES, A., IPPEN, E. P., SHANK, C. V.: (1971a), Appl. Phys. Letters **19**, 258.

DIENES, A., IPPEN, E. P., SHANK, C. V.: (1971b), unpublished.

DIENES, A., IPPEN, E. P., SHANK, C. V.: (1972), IEEE J. Quant. Electr. QE-**8**, 388.

DINEV, S. G., STAMENOV, K. V., TOMOV, I. V.: (1972), Opt. Commun. **5**, 419.

DREXHAGE, K. H.: (1971), unpublished results.

DREXHAGE, K. H.: (1972a), Design of Laser Dyes, VII. International Quantum Electronics Conference, Montreal, Canada.

DREXHAGE, K. H.: (1972b), unpublished results.

DREXHAGE, K. H.: (1973a), Laser Focus **9** (3), 35.

DREXHAGE, K. H.: (1973b), to be published.

DREXHAGE, K. H., HAWKS, G. H., REYNOLDS, G. A.: (1972), unpublished results.

DREXHAGE, K. H., REYNOLDS, G. A.: (1972), unpublished results.

DUNNING, F. B., STOKES, E. D., STEBBING, R. F.: (1972), Opt. Commun. **6**, 63.

DUNNING, F. B., TITTEL, F. K., STEBBING, R. F.: (1973), to be published.

DUTTON, L., KIHARA, T., McCRAY, J., THORNBER, J. P.: (1971), Biochim. and Biophys. Acta **226**, 81.

ECKHARDT, G.: (1966), IEEE J. Quant. Electr. QE-**2**, 1.

ECKHARDT, R. C., LEE, C. H.: (1969), Appl. Phys. Letters **15**, 425.

EINSTEIN, A.: (1910), Ann. Physik **33**, 1275.

EISENTHAL, K. B., DREXHAGE, K. H.: (1969), J. Chem. Phys. **51**, 5720.

ERDMANN, T. A., FIGGER, H., WALTHER, H.: (1972a), Opt. Commun. **6**, 166.

ERDMANN, T. A., FIGGER, H., HARTIG, W., SCHIEDER, R., WALTHER, H.: (1972b), Third International Conference on Atomic Physics, Boulder, Colo. Abstracts of Papers, pp. 68.

ERECINSKA, M., CHANCE, B.: (1972), Arch. Biochem. Biophys. 151, 304.

ERICKSON, L. E., SZABO, A.: (1971), Appl. Phys. Letters 18, 433.

FARIES, D. W., RICHARDS, P. L., SHEN, Y. R., YANG, K. H.: (1971), Phys. Rev. A3, 2148.

FERGUSON, J., MAU, A. W. H.: (1972a), Chem. Phys. Letters 14, 245.

FERGUSON, J., MAU, A. W. H.: (1972b), Chem. Phys. Letters 17, 543.

FERRAR, C. M.: (1969a), IEEE J. Quant. Electr. QE-5, 621.

FERRAR, C. M.: (1969b), IEEE J. Quant. Electr. QE-5, 550.

FLAMANT, P., MEYER, Y. H.: (1971), Appl. Phys. Letters 19, 491.

FLAMANT, P., MEYER, Y. H.: (1973), Opt. Commun. 7, 146.

FLECK, J. A.: (1968), J. Appl. Phys. 39, 3318.

FÖRSTER, TH.: (1951), Fluoreszenz organischer Verbindungen. Göttingen: Vandenhoeck and Ruprecht.

FÖRSTER, TH.: (1959), Discuss. Faraday Soc. 27, 7.

FÖRSTER, TH.: (1969), Angew. Chemie 81, 364; internat. ed. 8, 333.

FÖRSTER, TH., KÖNIG, E.: (1957), Z. Elektrochem. 61, 344.

FÖRSTER, TH., SELINGER, B.: (1964), Z. Naturforsch. 19a, 38.

FÖRSTERLING, H. D., KUHN, H.: (1971), Physikalische Chemie in Experimenten. Ein Praktikum, p. 373. Weinheim/Bergstr.: Verlag Chemie.

FORK, R. L., GERMAN, K. R.: (1972), Appl. Phys. Letters 20, 139.

FORK, R. L., GERMAN, K. R., CHANDROSS, E. A.: (1971), Appl. Phys. Letters 18, 139.

FORK, R. L., KAPLAN, Z.: (1972), Appl. Phys. Letters 20, 472.

FOUCHE, D. G., CHANG, R. K.: (1972), Phys. Rev. Letters 29, 536.

FURUMOTO, H. W., CECCON, H. L.: (1969a), Appl. Optics 8, 1613.

FURUMOTO, H. W., CECCON, H. L.: (1969b), J. Appl. Phys. 40, 4204.

FURUMOTO, H. W., CECCON, H. L.: (1970), IEEE J. Quant. Electr. QE-6, 262.

GABEL, C., HERCHER, M.: (1972), IEEE J. Quant. Electr. QE-8, 850.

GACOIN, P., FLAMANT, P.: (1972), Opt. Commun. 5, 351.

GASSMAN, M. H., WEBER, H.: (1971), Opto-Electr. 3, 177.

GIBBS, W. E. K., KELLOG, H. A.: (1968), IEEE J. Quant. Electr. QE-4, 293.

GIBSON, A. J.: (1969), J. Sci. Instr. 2, 802.

GILSON, T.: (1970), Laser Raman Spectroscopy. New York: Wiley.

GIORDMAINE, J. A., RENTZEPIS, P. M., SHAPIRO, S. L., WECHT, K. W.: (1967), Appl. Phys. Letters 11, 216.

GLENN, W. H., BRIENZA, M. J., DeMARIA, A. J.: (1968), Appl. Phys. Letters 12, 54.

GOLDMAN, L., ROCKWELL, R. J., Jr.: (1971), Lasers in Medicine. New York: Gordon and Breach.

GORNIK, W., KAISER, D., LANGE, W., LUTHER, J., SCHULZ, H. H.: (1972), Opt. Commun. 6, 327.

GRAMS, G. W., WYMAN, C. M.: (1972), J. Appl. Meteorology 11, 1108.

GREGG, D. W., QUERRY, M. R., MARLING, J. B., THOMAS, S. J., DOBLER, C. V., DAVIES, N. J., BELEW, J. F.: (1970), IEEE J. Quant. Electr. QE-6, 270.

GREGG, D. W., THOMAS, S. J.: (1969), IEEE J. Quant. Electr. QE-5, 302.

GRISCHKOWSKY, D.: (1973), Phys. Rev., A7, 2096.

GRISCHKOWSKY, D., ARMSTRONG, J. A.: (1972), Phys. Rev. A6, 1566.

GRONAU, B., LIPPERT, E., RAPP, W.: (1972), Ber. Bunsenges. Phys. Chem. 76, 432.

GUTFELD, R. J. VON, WELBER, B., TYNAN, E. E.: (1970), IEEE J. Quant. Electr. QE-6, 532.

HÄNSCH, T. W.: (1972), Appl. Opt. 11, 895.

HÄNSCH, T. W.: (1973), Specroscopy with Tunable Lasers, Third Intern. Conference on Atomic Physics, Boulder, Colo. pp. 579. SMITH, S. J., WALTERS, G. K. (Eds.). New York: Plenum Press.

HÄNSCH, T. W., SCHAWLOW, A. L.: (1970), Bull. Am. Phys. Soc. **15**, 1638.

HÄNSCH, T. W., PERNIER, M., SCHAWLOW, A. L.: (1971a), IEEE J. Quant. Electr. QE-**7**, 45.

HÄNSCH, T. W., VARSANYI, F., SCHAWLOW, A. L.: (1971b), Appl. Phys. Letters **18**, 108.

HÄNSCH, T. W., SHAHIN, I. S., SCHAWLOW, A. L.: (1971c), Phys. Rev. Letters **27**, 707.

HÄNSCH, T. W., SHAHIN, I. S., SCHAWLOW, A. L.: (1972a), Nature **235**, 63.

HÄNSCH, T. W., SCHAWLOW, A. L., TOSCHEK, P.: (1972b), IEEE J. Quant. Electr. QE-**8**, 802.

HAMADANI, S. M., MAGYAR, G.: (1971), Opt. Commun. **4**, 310.

HAMMOND, P. R., HUGHES, R. S.: (1971), Nature Phys. Sci. **231**, 59.

HANNA, D. C., SMITH, R. C., STANLEY, C. R.: (1971), Opt. Commun. **4**, 300.

HAPPER, W.: (1972), Rev. Mod. Phys. **44**, 169.

HAROCHE, S., PAISNER, J. A., SCHAWLOW, A. L.: (1973), Phys. Rev. Letters, accepted for publication.

HARRIS, S. E.: (1966), Proc. IEEE **54**, 1401.

HARRIS, S. E.: (1969), Proc. IEEE **57**, 2096.

HARTIG, W., WALTHER, H.: (1973), Appl. Phys. **1**, 171.

HAUSSER, K. W., KUHN, R., KUHN, E.: (1935), Z. Physik. Chemie B **29**, 417.

HELD, M.: (1968), Paper presented at the 8th Conference on High Speed Photography, Stockholm.

HERCHER, M., PIKE, H. A.: (1971), Opt. Commun. **3**, 65.

HERCHER, M., PIKE, H. A.: (1971a), Opt. Commun. **3**, 346.

HERCHER, M., PIKE, H. A.: (1971b), IEEE J. Quant. Electr. QE-**7**, 473.

HERCHER, M., SNAVELY, B. B.: (1972), Conference on Coherence and Quantum Optics, Rochester.

HERCULES, D. M.: (1964), Science **145**, 808.

HILDEBRAND, B. P., HAINES, K. A.: (1967), J. Opt. Soc. Am. **57**, 155.

HILL, K. O., WATANABE, A.: (1972), Opt. Commun. **5**, 389.

HINKLEY, E. D., KELLEY, P. L.: (1971), Science **171**, 635.

HIRTH, A., FAURE, J., SCHOENENBERGER, R.: (1972), C.R.Acad. Sc. Paris **274**B, 747.

HOLZRICHTER, J. F., McFARLANE, R. M., SCHAWLOW, A. L.: (1971), Phys. Rev. Letters **26**, 652.

HOROBIN, R. W., MURGATROYD, L. B.: (1969), Stain Technol. **44**, 297.

HOTOP, H., LINEBERGER, W. C.: (1973), J. Chem. Phys., to be published.

HOTOP, H., LINEBERGER, W. C.: (1973), Phys. Rev. A, to be published.

HUTH, B. G.: (1970), Appl. Phys. Letters **16**, 185.

HUTH, B. G., FARMER, G. I.: (1968), IEEE J. Quant. Electr. QE-**4**, 427.

HUTH, B. G., FARMER, G. I., KAGAN, M. R.: (1969), J. Appl. Phys. **40**, 5145.

IPPEN, E. P.: (1970), Appl. Phys. Letters **16**, 303.

IPPEN, E. P., SHANK, C. V., DIENES, A.: (1971), IEEE J. Quant. Electr. QE-**7**, 178.

IPPEN, E. P., SHANK, C. V.: (1972a), Appl. Phys. Letters **21**, 301.

IPPEN, E. P., SHANK, C. V., DIENES, A.: (1972b), Appl. Phys. Letters **21**, 348.

IPPEN, E. P., SHANK, C. V.: (1972c), unpublished.

ITZKAN, I., CUNNINGHAM, F. W.: (1972), IEEE J. Quant. Electr. QE-**8**, 101.

JAKOBI, H., KUHN, H.: (1962), Z. Elektrochem. Ber. Bunsenges. Phys. Chem. **66**, 46.

JENNINGS, D. A., VARGA, A. V.: (1971), J. Appl. Phys. **42**, 5171.

JOHNSON, S. E., BROIDA, H. P.: (1970), Bull. Am. Phys. Soc. **15**, 1630.

JOHNSON, F. M., SWAGEL, M. W.: (1971), Appl. Optics **10**, 1624.

JOURNAL OF CURRENT LASER ABSTRACTS, INSTITUTE FOR LASER DOCUMENTATION, FELTON, CALIF., SINCE 1964.

KAGAN, M. R., FARMER, G. I., HUTH, B. G.: (1968), Laser Focus **4** (9), 26.

KALLARD, Z.: (1969), Holography-state of the Art Review. New York: Optosonics Press.

KAMINOW, I. P., WEBER, H. P., CHANDROSS, E. A.: (1971), Appl. Phys. Letters **18**, 497.

KAMINOW, I. P., STULZ, L. W., CHANDROSS, E. A., PRYDE, C. A.: (1972), Appl. Optics **11**, 1563.

KARL, N.: (1972), Phys. Stat. Sol. (a) **13**, 651.

KATO, D., SATO, T.: (1972), Opt. Commun. **5**, 134.

KELLER, R. A.: (1970), IEEE J. Quant. Electr. **6**, 411.

KELLER, R. A., ZALEWSKI, E. F., PETERSON, N. C.: (1972), J. Opt. Soc. Am. **62**, 319.

KELLOGG, R. E.: (1970), J. Luminesc. **1**, **2**, 435.

KLAUDER, J. R., DUGUAY, M. A., GIORDMAINE, J. A., SHAPIRO, S. L.: (1968), Appl. Phys. Letters **13**, 174.

KLEIN, M. B.: (1972), Opt. Commun. **5**, 114.

KNIBBE, H., REHM, D., WELLER, A.: (1968), Ber. Bunsenges. Phys. Chem. **72**, 257.

KOGELNIK, H., YARIV, A.: (1964), Proc. IEEE **52**, 165.

KOGELNIK, H.: (1965), Bell Syst. Tech. J. **44**, 455.

KOGELNIK, H., LI, T.: (1966), Appl. Optics **5**, 1550.

KOGELNIK, H., SHANK, C. V., SOSNOWSKI, T. P., DIENES, A.: (1970), Appl. Phys. Letters **16**, 499.

KOGELNIK, H., SHANK, C. V.: (1971), Appl. Phys. Letters **18**, 152.

KOGELNIK, H., SHANK, C. V.: (1972a), J. Appl. Phys. **43**, 2327.

KOGELNIK, H., IPPEN, E. P., DIENES, A., SHANK, C. V.: (1972b), IEEE J. Quant. Electr. QE-**8**, 373.

KOHLMANNSPERGER, J.: (1969), Z. Naturforsch. **24**a, 1547.

KOHN, R. L., SHANK, C. V., IPPEN, E. P., DIENES, A.: (1971), Opt. Commun. **3**, 177.

KOROLEV, F., BAKHRAMOV, S. A., ODINTSOV, V. I.: (1970), JETP Letters **11**, 90.

KOTZUBANOV, V. D., NABOIKIN, YU. V., OGURTSOVA, L. A., PODGORNYI, A. P., POKROVSKAYA, F. S.: (1968a), Opt. Spectrosc. **25**, 406; Russ.: **25**, 727.

KOTZUBANOV, V. D., MALKES, L. YA., NABOIKIN, YU. V., OGURTSOVA, L. A., PODGORNYI, A. P., POKROVSKAYA, F. S.: (1968b), Bull. Acad. Sci. USSR, Phys. Ser. **32**, 1357; Rus.: **32**, 1466.

KRYUKOV, P. G., LETOKHOV, V. S.: (1970), Sov. Phys. Uspekhi **12**, 641.

KUHL, J., MAROWSKY, G.: (1971), Opt. Commun. **4**, 125.

KUHL, J., SPITSCHAN, H.: (1972), Opt. Commun. **5**, 382.

KUHN, H.: (1955), Chimia **9**, 237.

KUHN, H.: (1958—59), in: Progress in the Chemistry of Organic Natural Products, Vol. 16, pp. 17. (ZECHMEISTER, D. L., Ed.) Wien: Springer Verlag.

KUIZENGA, D. J., SIEGMAN, A. E.: (1970), IEEE J. Quant. Electr. QE-**6**, 694.

KUIZENGA, D. J.: (1971), Appl. Phys. Letters **19**, 260.

KUNG, A. H., YOUNG, J. F., HARRIS, S. E.: (1973a), Appl. Phys. Letters **22**, 301.

KUNG, A. H., YOUNG, J. F., BJORKLUND, G. C., HARRIS, S. E.: (1973b), to be published.

LABHART, H.: (1964), Helv. Chim. Acta **47**, 2279.

LAMM, M. E., NEVILLE, D. M., Jr.: (1965), J. Phys. Chem. **69**, 3872.

LARSSON, R., NORDÉN, B.: (1970), Acta Chem. Scand. **24**, 2583.

LEHMANN, M.: (1970), Holography, technique and practice. London, New York: Focal Press.

LEMPICKI, A., SAMELSON, H.: (1966),. In: Lasers, Vol. 1, pp. 181, LEVINE, A. K. (Ed.). New York: Marcel Dekker.

LEONHARDT, H., WELLER, A.: (1962), in: Luminescence of Organic and Inorganic Materials, p. 74. KALLMANN, H. P., SPRUCH, G. M. (Eds.). New York: Wiley.

LETOKHOV, V. S.: (1973), The use of lasers to control selective chemical reactions, Science, to be published.

LEVENSON, M. D., FLYTZANIS, C., BLOEMBERGEN, N.: (1972), Phys. Rev. B**6**, 3962.

LEVSHIN, L. V., GORSHKOV, V. K.: (1961), Opt. Spectrosc. **10**, 401; Russ.: **10**, 759.

LIDHOLT, L. R., WLADIMIROFF, W. W.: (1970), Opto-Electronics **2**, 21.

LINEBERGER, W. C.: (1972), Proc. of the Summer Research Conf. on Theoretical Chemistry, Boulder, Colo. New York: Wiley in press.

LINEBERGER, W. C., PATTERSON, T. A.: (1972), Chem. Phys. Letters **13**, 40.

LINEBERGER, W. C., WOODWARD, B. W.: (1970), Phys. Rev. Letters **25**, 424.

LIPPERT, E.: (1957), Z. Elektrochem. **61**, 962.

MACK, M. E.: (1969 a), Appl. Phys. Letters **15**, 81.

MACK, M. E.: (1969 b), Appl. Phys. Letters **15**, 166.

MAEDA, M., MIYAZOE, Y.: (1972), Japan. J. Appl. Phys. **11**, 692.

MAGYAR, G., SCHNEIDER-MUNTAU, H. J.: (1972), Appl. Phys. Letters **20**, 406.

MARLING, J. B., GREGG, D. W., THOMAS, S. J.: (1970 a), IEEE J. Quant. Electr. QE-**6**, 570.

MARLING, J. B., GREGG, D. W., WOOD, L.: (1970 b), Appl. Phys. Letters **17**, 527.

MARLING, J. B., WOOD, L. L., GREGG, D. W.: (1971), IEEE J. Quant. Electr. QE-**7**, 498.

MAROWSKY, G., RINGWELSKI, L., SCHÄFER, F. P.: (1972), Z. Naturforsch. **27a**, 711.

MAROWSKY, G.: (1973 a), IEEE J. Quant. Electr., QE-**9**, 245.

MAROWSKY, G.: (1973 b), Rev. Sci. Instr. **44**, 890.

MARTH, K.: (1967), Diplomarbeit, Universität Marburg, unpublished.

MAYER, S. W., KWOK, M. A., GROSS, R. W. F., SPENCER, D. J.: (1970), Appl. Phys. Lett. **17**, 516.

MCCALL, S. L., HAHN, E. L.: (1967), Phys. Rev. Letters **18**, 908.

MCCRAY, J. A.: (1972), Biochem. Biophys. Res. Commun. **47**, 187.

MCFARLAND, B. B.: (1967), Appl. Phys. Letters **10**, 208.

MCILRATH, T. J.: (1969), Appl. Phys. Letters **15**, 41.

MCNICE, G. T.: (1972), Appl. Optics **11**, 699.

MEASURES, R. M.: (1968), J. Appl. Phys. **39**, 5232.

MELFI, S. H.: (1972), Appl. Optics **11**, 1605.

MELHUISH, W. H.: (1962), J. Opt. Soc. Am. **52**, 1256.

MELNGAILIS, I.: (1972), IEEE Transact. Geosci. Electron. GE-**10**, 7.

MELTZER, D. W., GOLDBERG, L. S.: (1972), Opt. Commun. **5**, 209.

MENNICKE, H.: (1971), Phys. Letters **36** A, 127.

MILES, R. B., HARRIS, S. E.: (1973), IEEE J. Quant. Electr. QE-**9**, 470.

MIYAZOE, Y., MAEDA, M.: (1968), Appl. Phys. Letters **12**, 206.

MIYAZOE, Y., MAEDA, M.: (1970), Opto-Electronics **2**, 227.

MOORE, C. B.: (1973), to be published.

MORANTZ, D. J.: (1963), Proc. of the Symposium on Optical Masers, p.491. Brooklyn: Polytechnic Press.

MORANTZ, D. J., WHITE, B. G., WRIGHT, A. J. C.: (1962), Phys. Rev. Letters **8**, 23.

MORROW, T., QUINN, M.: (1973), to be published.

MÜLLER, A.: (1968), Z. Naturforsch. **23** a, 946.

MÜLLER, A., PFLÜGER, E.: (1968), Chem. Phys. Letters **2**, 155.

MÜLLER, A., SOMMER, U.: (1969), Ber. Bunsenges. **73**, 819.

MYER, J. A., ITZKAN, I., KIERSTEAD, E.: (1970), Nature **225**, 544.

MYERS, S. A.: (1971), Opt. Commun. **4**, 187.

NABOIKIN, YU. V., OGURTSOVA, L. A., PODGORNYI, A. P., POKROVSKAYA, F. S., GRIGORYEVA, V. I., KRASOVITSKII, B. M., KUTSYNA, L. M., TISHCHENKO, V. G.: (1970), Opt. Spectrosc. **28**, 528; Russ.: **28**, 974.

NAKASHIMA, M., SOUSA, J. A., CLAPP, R. C.: (1972), Nature Phys. Sci. **235**, 16.

NAKATO, Y., YAMAMOTO, N., TSUBOMURA, H.: (1968), Chem. Phys. Letters **2**, 57.

NEPORENT, B. S., SHILOV, V. B.: (1971), Opt. Spectrosc. **30**, 576; Russ.: **30**, 1074.

NEW, G. H. C.: (1972), Opt. Commun. **6**, 188.

NIEUWENHUIJZEN, H.: (1970), Optics Technology **2**, 13 and **2**, 68.

NOUCHI, G.: (1969), J. Chim. Phys. **66**, 548.

NOVAK, J. R., WINDSOR, M. W.: (1967), J. Chem. Phys. **47**, 3075.

NOVAK, J. R., WINDSOR, M. W.: (1968), Proc. Roy. Soc. A **308**, 95.

O'NEILL, F.: (1972), Opt. Commun. **6**, 360.

OPOWER, H., KAISER, W.: (1966), Phys. Letters **21**, 638.

OSBORNE, A. D., PORTER, G.: (1965), Proc. Roy. Soc. A **284**, 9.

PAPPALARDO, R., SAMELSON, H., LEMPICKI, A.: (1970a), Appl. Phys. Letters **16**, 267.

PAPPALARDO, R., SAMELSON, H., LEMPICKI, A.: (1970b), IEEE J. Quant. Electr. QE-**6**, 716.

PARKER, C. A.: (1968), Photoluminescence of Solutions. Amsterdam: Elsevier.

PARKER, C. A., HATCHARD, C. G.: (1961), Transact. Faraday Soc. **57**, 1894.

PARSONS, F. G., CHANG, R. K.: (1971), Opt. Commun. **3**, 173.

PARSONS, F. G., CHEN, E. YI., CHANG, R. K.: (1971), Phys. Rev. Letters **27**, 1436.

PERSONOV, R. I., SOLOCHMOV, V. V.: (1967), Opt. Spectrosc. **23**, 317.

PETERSON, N. C., KURYLO, M. J., BRAUN, W., BASS, A. M., KELLER, R. A.: (1971), J. Opt. Soc. Am. **61**, 746.

PETERSON, O. G., SNAVELY, B. B.: (1968), Appl. Phys. Letters **12**, 238.

PETERSON, O. G., TUCCIO, S. A., SNAVELY, B. B.: (1970), Appl. Phys. Letters **17**, 245.

PETERSON, O. G., WEBB, J. P., McOLGIN, W. C., EBERLY, J. H.: (1971), J. Appl. Phys. **42**, 1917.

PETROFF, Y., YU, P. Y., SHEN, Y. R.: (1972), Phys. Rev. Letters **29**, 1558.

PICARD, R. H., SCHWEITZER, P.: (1970), Phys. Rev. A **1**, 1803.

PIKE, H. A.: (1971) Ph. D. thesis, University of Rochester, available from University Microfilms, Ann Arbor, Mich., USA.

POLE, R. V., MILLER, S. E., HARRIS, J. H., TIEN, P. K.: (1972), Appl. Optics **11**, 1675.

PORTER, G., WINDSOR, M. W.: (1958), Proc. Roy. Soc. A **245**, 238.

PRINGSHEIM, P.: (1949), Fluorescence and Phosphorescence. New York: Interscience.

RABINOWITCH, E., EPSTEIN, L. F.: (1941), J. Am. Chem. Soc. **63**, 69.

RAMETTE, R. W., SANDELL, E. B.: (1956), J. Am. Chem. Soc. **78**, 4872.

RAUTIAN, S. G., SOBEL'MANN, I. I.: (1961), Opt. Spectrosc. **10**, 65.

REBANE, K. K., KHIZHNYAKOV, V. V.: (1963), Opt. Spectrosc. **14**, 262.

REINHOLD, I., MAIER, M.: (1972), Opt. Commun. **5**, 31.

RENTZEPIS, P. M., MITSCHELE, C. J., SAXMAN, A. C.: (1970), Appl. Phys. Letters **17**, 122.

REYNOLDS, G. A., DREXHAGE, K. H.: (1972), unpublished results.

RINGWELSKI, L., SCHÄFER, F. P.: (1970), unpublished results.

ROBINSON, G. W., FROSCH, R. P.: (1963), J. Chem. Phys. **38**, 1187.

ROHATGI, K. K., MUKHOPADHYAY, A. K.: (1971), Photochem. Photobiol. **14**, 551.

ROHATGI, K. K., SINGHAL, G. S.: (1966), J. Phys. Chem. **70**, 1695.

ROSS, M. (Ed.): (1971), Laser Applications. New York: Academic Press.

ROWE, H. E., LI, T.: (1970), IEEE J. Quant. Electr. QE-**6**, 49.

RUBINOV, A. N., MOSTOVNIKOV, V. A.: (1967), J. Appl. Spectrosc. **7**, 223; Russ.: **7**, 327.

RUBINOV, A. N., MOSTOVNIKOV, V. A.: (1968), Bull. Acad. Sci. USSR, Phys. Ser. **32**, 1348; Russ.: **32**, 1456.

RUFF, G. A., HERCHER, M., PIKE, H. A.: (1971), Bull. Am. Phys. Soc. **16**, 1340.

RUNGE, P. K.: (1971), Opt. Commun. **4**, 195.

RUNGE, P. K.: (1972), Opt. Commun. **5**, 311.

RUNGE, P. K., ROSENBERG, R.: (1972), IEEE J. Quant. Electr. QE-**8**, 910.

SACKETT, P. B., YARDLEY, J. T.: (1970), Chem. Phys. Letters **6**, 323.

SAKURAI, K., CAPELLE, G.: (1970), J. Chem. Phys. **53**, 3764.

SAKURAI, K., CAPELLE, G., BROIDA, H. P.: (1971), J. Chem. Phys. **54**, 1220.

SANDFORD, M. C. W., GIBSON, A. J.: (1970), J. Atm. Terr. Phys. **32**, 1423.

SCHÄFER, F. P.: (1968), Invited Paper at Internat. Quant. Electr. Conf., Miami, Fla., USA.

SCHÄFER, F. P.: (1969), Conference on Nonlinear Optics, Belfast, UK.

SCHÄFER, F. P.: (1970), Angew. Chem. **82**, 25; internat. ed. **9**, 9.

SCHÄFER, F. P., RINGWELSKI, L.: (1973), Z. Naturforsch. **28**a, 792.

SCHÄFER, F. P., SCHMIDT, W., VOLZE, J.: (1966), Appl. Phys. Letters **9**, 306.

SCHÄFER, F. P., SCHMIDT, W., MARTH, K.: (1967), Phys. Letters **24**A, 280.

SCHÄFER, F. P., SCHMIDT, W., VOLZE, J., MARTH, K.: (1968), Ber. Bunsenges. phys. Chemie **72**, 328.

SCHÄFER, F. P., MÜLLER, H.: (1971), Opt. Commun. **2**, 407.

SCHAPPERT, G. T., BILLMAN, K. W., BURNHAM, D. C.: (1968), Appl. Phys. Letters **13**, 124.

SCHEIBE, G.: (1941), Z. Elektrochem. **47**, 73.

SCHENCK, P., HILBORN, R. C., METCALF, H.: (1973), Bull. Am. Phys. Soc. **18**, 121.

SCHIEDER, R., WALTHER, H., WOSTE, L.: (1972), Opt. Commun. **5**, 337.

SCHINKE, D. P., SMITH, R. G., SPENCER, E. G., GALVIN, M. F.: (1972), Appl. Phys. Letters **21**, 494.

SCHMIDT, W.: (1970), Laser **2**, 47.

SCHMIDT, W., SCHÄFER, F. P.: (1967), Z. Naturforsch. **22**a, 1563.

SCHMIDT, W., SCHÄFER, F. P.: (1968), Phys. Letters **26**A, 558.

SCHMIDT, W., FERCHER, A. F.: (1971), Opt. Commun. **3**, 363.

SCHMIDT, W., WITTEKINDT, N.: (1972), Appl. Phys. Letters **20**, 71.

SCHMIDT, W., VOGEL, A., PREUSSLER, D.: (1973), Appl. Phys. **1**, 103.

SCHMIDT, W., APPT. W.: (1973), Z. Naturforsch., to be published.

SCHRÖDER, W., LINS, E., WELLEGEHAUSEN, B., WELLING, H.: (1973), Appl. Phys. **1**, 343.

SCHUDA, F., HERCHER, M., STROUD, C. R.: (1973), Opt. Commun., to be published.

SCHULER, C. J., PIKE, C. T., MIRANDA, H. A.: (1971), Appl. Optics **10**, 1689.

SELWYN, J. E., STEINFELD, J. I.: (1972), J. Phys. Chem. **76**, 762.

SEVCHENKO, A. N., PIKULIK, L. G., GLADCHENKO, L. F., DAS'KO, A. D.: (1968a), J. Appl. Spectrosc. **8**, 556; Russ.: **8**, 935.

SEVCHENKO, A. N., KOVALEV, A. A., PILIPOVICH, V. A., RAZVIN, YU. V.: (1968b), Sov. Phys.-Doklady **13**, 226.

SHANK, C. V., DIENES, A., TROZZOLO, A. M., MYER, J. A.: (1970a), Appl. Phys. Letters **16**, 405.

SHANK, C. V., DIENES, A., SILFVAST, W. T.: (1970b), Appl. Phys. Letters **17**, 307.

SHANK, C. V., BJORKHOLM, J. E., KOGELNIK, H.: (1971), Appl. Phys. Letters **18**, 395.

SHANK, C. V., IPPEN, E. P., DIENES, A.: (1972), Digest of Tech. Papers, VII International Quant. Electr. Conf., Montreal.

SHAPIRO, S. L., DUDUAY, M. A.: (1969), Phys. Letters **28**A, 698.

SHIMODA, K.: (1973), Appl. Phys. **1**, 77.

SHPOL'SKII, E. V.: (1962), Usp. Fiz. Nauk **77**, 321 and **80**, 255.

SIEGMAN, A. E.: (1971), Appl. Optics **10**, A 38.

SIEGMAN, A. E., PHILLION, D. W., KUIZENGA, D. J.: (1972), Appl. Phys. Letters **21**, 345.

SINGH, S.: (1971), Stimulated Raman Scattering, in CRC Handbook of Lasers, Pressley, R. J. ed. Cleveland: CRC.

SMITH, H. M.: (1969), Principles of Holography. New York: Wiley.

SMITH, P. W.: (1970), Proc. IEEE **58**, 1342.

SMITH, W. V., SOROKIN, P. P.: (1966), The Laser, p. 74. New York: McGraw Hill.

SMITH, W. V.: (1971), Laser Applications. Dedham: Artech house.

SMOLSKAYA, T. I., RUBINOV, A. N.: (1971), Opt. Spectrosc. **31**, 235.

SNAVELY, B. B.: (1969), Proc. IEEE **57**, 1374.

SNAVELY, B. B., PETERSON, O. G., REITHEL, R. F.: (1967), Appl. Phys. Letters **11**, 275.

SNAVELY, B. B., PETERSON, O. G.: (1968), IEEE J. Quant. Electr. QE-4, 540.

SNAVELY, B. B., SCHÄFER, F. P.: (1969), Phys. Letters **28**A, 728.

SOEP, B.: (1970), Opt. Commun. **1**, 433.

SOEP, B., KELLMANN, A., MARTIN, M., LINDQVIST, L.: (1972), Chem. Phys. Letters **13**, 241.

SOFFER, B. H., McFARLAND, B. B.: (1967), Appl. Phys. Letters **10**, 266.

SOFFER, B. H., LINN, J. W.: (1968), J. Appl. Phys. **39**, 5859.

SOREM, M. S., SCHAWLOW, A. L.: (1972), Opt. Commun. **5**, 148.

SOROKIN, P. P.: (1969), Sci. Am. **220**, (2) 30.

SOROKIN, P. P., LANKARD, J. R.: (1966), IBM J. Res. Develop. **10**, 162.

SOROKIN, P. P., CULVER, W. H., HAMMOND, E. C., LANKARD, J. R.: (1966a), IBM J. Res. Develop. **10**, 401.

SOROKIN, P. P., CULVER; W. H., HAMMOND, E. C., LANKARD, J. R.: (1966b), IBM J. Res. Develop. **10**, 428.

SOROKIN, P. P., LANKARD, J. R., HAMMOND, E. C., MORUZZI, V. L.: (1967), IBM J. Res. Develop. **11**, 130.

SOROKIN, P. P., LANKARD, J. R.: (1967), IBM J. Res. Develop. **11**, 148.

SOROKIN, P. P., LANKARD, J. R., MORUZZI, V. L., HAMMOND, E. C.: (1968), J. Chem. Phys. **48**, 4726.

SOROKIN, P. P., LANKARD, J. R., MORUZZI, V. L.: (1969a), Appl. Phys. Letters **15**, 179.

SOROKIN, P. P., LANKARD, J. R.: (1969b), Phys. Rev. **186**, 342.

SOROKIN, P. P., LANKARD, J. R.: (1971), J. Chem. Phys. **54**, 2184.

SOROKIN, P. P., WYNNE, J. L., LANKARD, J. R.: (1973), Appl. Phys. Letters **22**, 342.

SPAETH, M. L., BORTFELD, D. P.: (1966), Appl. Phys. Letters **9**, 179.

SRINIVASAN, R.: (1969), IEEE J. Quant. Electr. QE-5, 552.

STEBBING, R. F., DUNNING, F. B.: (1973), to be published.

STEPANOV, B. I., RUBINOV, A. N., MOSTOVNIKOV, V. A.: (1967a), J. Appl. Spectry. **7**, 116.

STEPANOV, B. I., RUBINOV, A. N., MOSTOVNIKOV, V. A.: (1967b), JETP Letters **5**, 117; Russ.: **5**, 144.

STEPANOV, B. I., RUBINOV, A. N.: (1968), Sov. Phys. Usp. **11**, 304.

STOCKMAN, D. L.: (1964), Proc. of the ONR Conf. on Organic Lasers, Doc. no. AD 447468, Defence Documentation Center for Scientific and Technical Information, Cameron Station, Alexandria, Va., USA.

STOCKMAN, D. L., MALLORY, W. R., TITTEL, F. K.: (1964), Proc. IEEE **52**, 318.

STRICKLER, S. J., BERG, R. A.: (1962), J. Chem. Phys. **37**, 814.

STROME, F. C., Jr., WEBB, J. P.: (1971), Appl. Optics **10**, 1348.

STROME, F. C., Jr., TUCCIO, S. A.: (1971), Opt. Commun. **4**, 58.

STROME, F. C., Jr.: (1972), IEEE J. Quant. Electr. QE-8, 98.

SUSCHINSKII, M. M.: (1972), Raman Spectra of Molecules and Crystals. New York: Wiley.

SZABO, A., ERICKSON, L. E.: (1972), Opt. Commun. **5**, 287.

TAYLOR, D. J., HARRIS, S. E., NIEH, S. K. T., HÄNSCH, T. W.: (1971), Appl. Phys. Letters **19**, 269.

TERENIN, A. N., ERMOLAEV, V. L.: (1956), Usp. Fiz. Nauk **58**, 37.

TERHUNE, R. W., MAKER, P. D.: (1968), Lasers, Vol. 2, LEVINE, A. K. (Ed.). New York: Dekker.

THRASH, R. J., WEYSSENHOFF, H. VON, SHIRK, J. S.: (1971), J. Chem. Phys. **55**, 4659.

TIEN, P. K.: (1971), Appl. Optics **10**, 2395.

TIFFANY, W. B., MOOS, H. W., SCHAWLOW, A. L.: (1967), Science **157**, 40.

TOBIN, M. C.: (1971), Chemical Analysis, Vol. 35. New York: Wiley.

TOMIYASU, K.: (1968), The Laser Literature, an Annotated Guide. New York: Plenum Press.

TOPP, M. R., RENTZEPIS, P. M.: (1971), Phys. Rev. **3**A, 358.

TUCCIO, S. A., STROME, F. C., Jr.: (1972), Appl. Optics **11**, 64.

TUCCIO, S. A., DREXHAGE, K. H., REYNOLDS, G. A.: (1973), Opt. Commun **7**, 248.

TUREK, C. A., YARDLEY, J. T.: (1971), IEEE J. Quant. Electr. QE-7, 102.

TURRO, N. J.: (1965), Molecular Photochemistry. New York: Benjamin.

VARGA, P., KRYUKOV, P. G., KUPRISHOV, V. F., SENATSKII, YU. V.: (1968), JETP Letters **8**, 307; Russ.: **8**, 501.

VARGA, P., KRYUKOV, P. G., KUPRISHOV, V. F., SENATSKII, YU. V.: (1969), Opt. Spectrosc. **26**, 545.

VARSANYI, F.: (1971), Appl. Phys. Letters **19**, 169.

VIKTOROVA, E. N., GOFMAN, I. A.: (1965), Russ. J. Phys. Chem. **39**, 1416; Russ.: **39**, 2643.

VOLZE, J.: (1969), Dissertation, Marburg.

VREHEN, Q. H. F.: (1971), Opt. Commun. **3**, 144.

VREHEN, Q. H. F., BREIMER, A. J.: (1972), Opt. Commun. **4**, 416.

WALLACE, R. W.: (1971), Opt. Commun. **4**, 316.

WALLACE, R. W.: (1972), IEEE J. Quant. Electr. QE-**8**, 819.

WALTHER, H., HALL, J. L.: (1970), Appl. Phys. Letters **17**, 239.

WEBB, J. P., MCCOLGIN, W. C., PETERSON, O. G., STOCKMAN, D. L., EBERLY, J. H.: (1970), J. Chem. Phys. **53**, 4227.

WEBER, H. P., DANIELMEYER, H. G.: (1970), Phys. Rev. A**2**, 2074.

WEBER, H. P., ULRICH, R.: (1971), Appl. Phys. Letters **19**, 38.

WEBER, M. J., BASS, M.: (1969), IEEE J. Quant. Electr. QE-**5**, 175.

WEHRY, E. L.: (1967), Structural and environmental factors in fluorescence, in: Fluorescence; Theory, Instrumentation, and Practice, p. 37. GUILBAULT, G. G. (Ed.). New York: Dekker.

WELLER, A.: (1958), Z. phys. Chemie NF **18**, 163.

WIEDER, I.: (1972), Appl. Phys. Letters **21**, 318.

WINEFORDNER, J. D., ELSER, R. C.: (1971), Anal. Chem. **43**, 24 A.

WINEFORDNER, J. D., VICKERS, T. J.: (1972), Anal. Chem. **42**, 206 R.

WYNNE, J. J.: (1972), Phys. Rev. Letters **29**, 650.

WOLBARSHT, M. L., (Ed.): (1971), Laser Applications in Medicine and Biology. New York: Plenum Press.

YAMAGUCHI, G., ENDO, F., MURAKAWA, S., OKAMURA, S., YAMANAKA, C.: (1968), Japan. J. Appl. Phys. **7**, 179.

YAMAGUCHI, G., MURAKAWA, S., TANAKA, H., YAMANAKA, C.: (1969), Japan. J. Appl. Phys. **8**, 1265.

YARIV, A.: (1967), Quantum Electronics, Chap. 15. New York: Wiley.

YEUNG, E. S., MOORE, C. B.: (1972), Appl. Phys. Letters **21**, 109.

YGUERABIDE, J.: (1968), Rev. Sci. Instr. **39**, 1048.

YOUNG, J. F., BJORKLUND, G. C., KUNG, A. H., MILES, R. B., HARRIS, S. E.: (1971), Phys. Rev. Letters **27**, 1551.

YU, P. Y., SHEN, Y. R.: (1972), Phys. Rev. Letters **29**, 468.

ZANKER, V., MIETHKE, E.: (1957a), Z. Phys. Chem. NF **12**, 13.

ZANKER, V., MIETHKE, E.: (1957b), Z. Naturforsch. **12**a, 385.

ZEIDLER, G.: (1971), J. Appl. Phys. **42**, 884.

ZELENKA, J. S., VARMER, J. R.: (1968), Appl. Opt. **7**, 2107.

ZORY, P.: (1972), Paper presented at the OSA Topical Meeting on Integrated Optics, Las Vegas, Nev., USA.

Additional References with Titles

AHMED, S. A.: (1973), Molecular air pollution monitoring by dye laser measurement of differential absorption of atmospheric elastic backscattering. Appl. Optics **12**, 901.

ALFANO, R. R., SHAPIRO, S. L., YU, W.: (1973), Effect of soap on the fluorescent lifetime and quantum yield of rhodamine 6 G in water. Opt. Commun. **7**, 191.

ANLIKER, P., GASSMANN, M., WEBER, H.: (1972), 12 joules rhodamine 6 G laser. Opt. Commun. **5**, 137.

ARTHURS, E. G., BRADLEY, D. J., RODDIE, A. G.: (1973), Buildup of picosecond pulse generation in passively mode-locked rhodamine dye lasers. Appl. Phys. Letters **23**, 88.

ARTHURS, E. G., BRADLEY, D. J., RODDIE, A. G.: (1973), Photoisomer generation and absorption relaxation in the mode-locking dye 3,3'-diethyloxadicarbocyanine iodide. Opt. Commun., to be published.

AUSTON, D. H., GLASS, A. M., LEFUR, P.: (1973), Tunable far infrared generation by difference frequency mixing of dye lasers in reduced (black) lithium niobate. Appl. Phys. Letters **23**, 47.

BASTING, D., SCHÄFER, F. P., STEYER, B.: (1974), New laser dyes. Appl. Phys. **3**, (January issue).

BEER, D., WEBER, J.: (1972), Photobleaching of organic laser dyes. Opt. Commun. **5**, 307.

BJORKHOLM, J. E., SOSNOWSKI, T. P., SHANK, C. V.: (1973), Distributed-feedback lasers in optical waveguides deposited on anisotropic substrates. Appl. Phys. Letters **22**, 132.

BRITT, A. D., MONIZ, W. B.: (1972), The effect of pH on photobleaching of organic laser dyes. IEEE J. Quant. Electr. **8**, 913.

BRITT, A. D., MONIZ, W. B.: (1973), Reactivity of first-singlet excited xanthene laser dyes in solution. J. Org. Chem. **38**, 1057.

BUNKENBURG, J.: (1972), An 11 megawatt 6.8 joule flashlamp-pumped coaxial liquid dye laser. Rev. Sci. Instr. **43**, 1611.

BURLAMACCHI, P., PRATESI, R.: (1973), Waveguide superradiant dye laser. Appl. Phys. Letters **22**, 334.

CARLSTEN, J. L., McILRATH, T. J.: (1973), Observations of stimulated anti-Stokes radiation in barium vapor. J. Phys. B. **6**, L 80.

CHANG, M. S., BURLAMACCHI, P., HU, C., WHINNERY, J. R.: Light amplification in a thin film. Appl. Phys. Letters **20**, 313.

CHIN, S. L., LECLERC, L., BÉDARD, G.: (1972), New emission band from a superradiant traveling-wave rhodamine 6 G dye laser. Opt. Commun. **6**, 264.

DECKER, C. D., TITTEL, F. K.: (1973), Broadly tunable, narrow linewidth dye laser emission in the near infrared. Opt. Commun. **7**, 155.

DECKER, C. D., TITTEL, F. K.: (1973), Difference frequency generation by optical mixing of two dye lasers in proustite. Opt. Commun. **8**, 244.

DEMPSTER, D. N., MORROW, T., QUINN, M. F.: (1973), Extinction coefficients for triplet-triplet absorption in ethanol solutions of anthracene, naphthalene, 2,5-diphenyloxazole, 7-diethylamino-4-methyl coumarin and 4-methyl-7-amino-carbostyril. J. Photochemistry **2**, 1.

DEMPSTER, D. N., MORROW, T., QUINN, M. F.: (1973), The photochemical characteristics of rhodamine 6 G — ethanol solutions. J. Photochemistry **2**, 25.

DEMPSTER, D. N., MORROW, T., RANKIN, R., THOMPSON, G. F.: (1972), Photochemical characteristics of cyanine dyes. J. C. S. Faraday II **68**, 1479.

DEZAUZIER, P., ERANIAN, A., DEWITTE, O.: (1973), Amplification competition in a double-cavity flash-pumped dye laser. Appl. Phys. Letters **22**, 664.

DIENES, A., JAIN, R. K., LIN, C.: (1973), Formation mechanisms in an excited-state-reaction dye laser. Appl. Phys. Letters **22**, 632.

DIENES, A., SHANK, C. V., KOHN, R. L.: (1973), Characteristics of the 4-methylumbel-liferone laser dye. IEEE J. Quant. Electr. **9**, 833.

DUNNING, F. B., STOKES, E. D.: (1972), The generation of tunable near IR radiation using a nitrogen laser pumped dye laser. Opt. Commun. **6**, 160.

DUONG, H. T., JACQUINOT, P., LIBERMAN, S., PICQUÉ, J. L., PINARD, J., VIALLE, J. L.: (1973), Optical resolution of the hyperfine structure of the Na D lines by atomic beam absorption from a cw tunable dye laser. Opt. Commun. **7**, 371.

ERANIAN, A., DEZAUZIER, P., DEWITTE, O.: (1973), 2-nsec pulses from double cavity dye laser. Opt. Commun. **7**, 150.

EWANIZKY, T. F., WRIGHT JR., R. H., THEISSING, H. H.: (1973), Shock-wave termination of laser action in coaxial flashlamp dye lasers. Appl. Phys. Letters **22**, 520.

FERRAR, C. M.: (1972), Vortex-confined pumping discharge in dye laser solution. Appl. Phys. Letters **20**, 419.

GALE, G. M.: (1973), A single mode flashlamp-pumped dye laser. Opt. Commun. **7**, 86.

GILLARD, P. G., FOLTZ, N. D., CHO, C. W.: (1973), Gain in DTTC-methanol solution as a function of pump power, wavelength, and time. Appl. Phys. Letters **23**, 325.

GREEN, J. M., HOHIMER, J. P., TITTEL, F. K.: (1973), Traveling-wave operation of a tunable cw dye laser. Opt. Commun. **7**, 349.

GREEN, J. M., HOHIMER, J. P., TITTEL, F. K.: (1973), A high resolution cw dye laser spectrometer. Submitted to Opt. Commun.

GRISCHKOWSKY, D., COURTENS, E., ARMSTRONG, J. A.: (1973), Observation of slow velocities and self-steepening of optical pulses described by adiabatic following. Proc. of the Laser Spectroscopy Conference, Vail, Col., June 1973, Plenum Press, to be published.

GORNIK, W., KAISER, D., LANGE, W., LUTHER, J., RADLOFF, H. H., SCHULZ, H. H.: (1973), Lifetime measurements using stepwise excitation by two pulsed dye lasers. Appl. Phys. **1**, 285.

HANNA, D. C., RAMPEL, V. V., SMITH, R. C.: (1973), Tunable infrared down-conversion in silver thiogallate. Opt. Commun. **8**, 151.

HIRTH, A., VOLLRATH, K., FAURE, J., LOUGNOT, D.: (1973), Flashlamp-excited dye lasers in the near infrared. Opt. Commun. **7**, 339.

IOFFE, I. S., SHAPIRO, A. L.: (1972), Mutual conversions of the colorless and colored forms of N,N'-substituted rhodamines. Zh. Org. Khim. **8**, 1765; Russ.: **8**, 1726.

ISSA, I. M., ISSA, R. M., GHONEIM, M. M.: (1972), Spectrophotometric studies on fluorescein derivatives in aqueous solutions. J. Phys. Chemie, Leipzig **250**, 161.

JACOBS, R. R., LEMPICKI, A., SAMELSON, H.: (1973), Efficient and damage-resistant tunable cw dye laser. J. Appl. Phys. **44**, 2775.

JACOBS, R. R., SAMELSON, H., LEMPICKI, A.: (1973), Losses in cw dye lasers. J. Appl. Phys. **44**, 263.

JACQUINOT, P., LIBERMAN, S., PICQUÉ, J. L., PINARD, J.: (1973), High resolution spectroscopic application of atomic beam deflection by resonant light. Opt. Commun. **8**, 163.

KINDT, T., LIPPERT, E., RAPP, W.: (1972), The influence of the excitation intensity on the laser spectra of acidified solutions of 4-methylumbelliferone. Z. Naturforsch. **27a**, 1371.

KLEIN, M. B., SHANK, C. V., DIENES, A.: (1973), Detection of small laser gains in a helium-selenium discharge using dye laser assisted oscillation. Opt. Commun. **7**, 178.

Kohn, R. L., Shank, C. V., Dienes, A.: (1973), Observation of inhomogeneity in the gain spectrum of a coumarin laser dye. Opt. Commun. **7**, 309.

Kuhl, J., Spitschan, H.: (1973), Flame-fluorescence detection of Mg, Ni, and Pb with a frequency-doubled dye laser excitation source. Opt. Commun. **7**, 256.

Kushida, T., Tanaka, Y.: (1972), Direct optical excitation into excited states of Cr^{3+} pairs in ruby. Solid State Commun. **11**, 1341.

Lange, W., Luther, J., Nottbeck, B., Schröder, H. W.: (1973), High resolution fluorescence spectroscopy by use of a cw dye laser. Opt. Commun. **8**, 157.

Letonzey, J. P., Sari, S. O.: (1973), Continuous pulse train dye laser using an open flowing passive absorber. Appl. Phys. Letters **23**, 311.

Lin, C., Gustafson, T. K., Dienes, A.: (1973), Superradiant picosecond laser emission from transversely pumped dye solutions. Opt. Commun. **8**, 210.

Loth, C., Meyer, Y. H.: (1973), Study of a 1-watt repetitive dye laser. Appl. Optics **12**, 123.

Magyar, G.: (1972), Frequency shift in a mode-selected dye laser. Opt. Commun. **6**, 388.

Marowsky, G.: (1973), Spectral narrowing in a dye laser with non-resonant feedback. Appl. Phys. **2**, 213.

Marowsky, G.: (1973), Gain narrowing studies with an organic dye laser. Submitted to J. Appl. Phys.

McIlrath, T. J., Carlsten, J. L.: (1973), Production of large number of atoms in a selected excited state by laser optical pumping. J. Phys. B **6**, to be published.

Morou, G., Drouin, B., Bergeron, M., Denariez-Roberge, M. M.: (1973), Kinetics of bleaching in polymethine cyanine dyes. IEEE J. Quant. Electr. **9**, 745.

Mory, S., Leupold, D., König, R.: (1972), On the mechanism of rhodamine 6 G fluorescence quenching. Opt. Commun. **6**, 394.

Nakashima, N., Mataga, N., Yamanaka, C., Ide, R., Misumi, S.: (1973), Intramolecular exciplex laser. Chem. Phys. Letters **18**, 386.

Neumann, G., Wieder, I.: (1972), Longitudinal excitation of a short cavity tunable dye laser by a nitrogen laser. Opt. Commun. **5**, 197.

Oka, Y., Kushida, T.: (1972), Resonance Raman scattering in CdS and ZnO by tunable dye laser. J. Phys. Soc. Japan **33**, 1372.

Oka, Y., Kushida, T., Murahashi, T., Koda, T.: (1973), Resonance Raman scattering in cuprous halides. Techn. Report of ISSP, Series A, No. 601, Tokyo.

Pavlopoulos, T. G.: (1973), Prediction of laser action properties of organic dyes from their structure and the polarization characteristics of their electronic transitions. IEEE J. Quant. Electr. **9**, 510.

Pavlopoulos, T. G.: (1973), Measurements of molar triplet extinction coefficients or organic molecules by means of cw laser excitation. J. Opt. Soc. Am. **63**, 180.

Peters, R. L. St., Taylor, D. J.: (1973), Face-pumped high-average-power low-distortion dye laser. Appl. Phys. Letters **23**, 90.

Picqué, J.-L., Vialle, J.-L.: (1972), Atomic-beam deflection and broadening by recoils due to photon absorption or emission. Opt. Commun. **5**, 402.

Pilloff, H. S.: (1972), Simultaneous two-wavelength selection in the N_2 laser-pumped dye laser. Appl. Phys. Letters **21**, 339.

Ricard, D., Lowdermilk, W. H., Ducuing, J.: (1972), Direct observation of vibrational relaxation of dye molecules in solution. Chem. Phys. Letters **16**, 617.

Rohatgi, K. K., Mukhopadhyay, A. K.: (1972), Hypochromism and exciton interaction in halofluorescein dyes. J. Phys. Chem. **76**, 3970.

Rohatgi, K. K., Mukhopadhyay, A. K.: (1972), Aggregation properties of anions of fluorescein and halofluorescein dyes. J. Indian Chem. Soc. **49**, 1311.

Rullière, C., Denariez-Roberge, M. M.: (1973), Effet laser dans quelques composés organiques entre 3700 et 5000 Å. Opt. Commun. **7**, 166.

RULLIÈRE, C., LAUGHREA, M., DENARIEZ-ROBERGE, M. M.: (1972), Action laser dans le
  perylène à 4730 Å. Opt. Commun. **6**, 407.
SCHENCK, P., HILLBORN, R. C., METCALF, H.: (1973), Time resolved fluorescence from Ba
  and Ca excited by a pulsed tunable dye laser. Phys. Rev. Letters **31**, 189.
SCHIMITSCHEK, E. J., TRIAS, J. A., TAYLOR, M., CELTO, J. E.: (1973), New improved laser
  dye for the blue-green spectral region. IEEE J. Quant. Electr. **9**, 781.
SHAH, J.: (1972), New output coupling scheme for cw dye lasers. Appl. Phys. Letters
  20, 479.
SHANK, C. V., EDIGHOFFER, J., DIENES, A., IPPEN, E. P.: (1973), Evidence for diffusion
  independent triplet quenching in the rhodamine 6 G ethylene glycol cw dye laser
  system. Opt. Commun. **7**, 176.
SHANK, C. V., KLEIN, M. B.: (1973), Frequency locking of a cw dye laser near atomic
  absorption lines in a gas discharge. Appl. Phys. Letters **23**, 156.
SIEGMAN, A. E.: (1972), Dispersive explanation of the spectral behavior of Runge's
  mode-locked dye laser. Opt. Commun. **5**, 200.
STEBBINGS, R. F., DUNNING, F. B., TITTEL, F. K., RUNDEL, R. D.: Photoionization of helium
  metastable atoms near threshold. Opt. Commun., to be published.
STOKES, E. D., DUNNING, F. B., STEBBINGS, R. F., WALTERS, G. K., RUNDEL, R. D.: (1972),
  A high efficiency dye laser tunable from the UV to the IR. Opt. Commun. **5**, 267.
SVANBERG, S.: (1973), Spectroscopy of highly excited levels in alkali atoms using a cw
  tunable dye laser. Proc. of the Laser Spectroscopy Conference, Vail, Col., June 1973,
  Plenum Press, to be published.
SVANBERG, S., TSEKERN, P., HAPPER, W.: (1973), Hyperfine-structure studies of highly
  excited D and F levels in alkali atoms using a cw tunable dye laser. Phys. Rev. Letters
  30, 817.
TOMIN, V. I., BUSHUK, B. A., RUBINOV, A. N.: (1972), Study of the gain curve and the
  triplet-triplet absorption in a rhodamine 6 G solution laser. Opt. Spectrosc. **32**, 527;
  Russ.: **32**, 983.
WEBER, J.: (1973), Study of the influence of triplet quencher on the photobleaching of
  rhodamine 6 G. Opt. Commun. **7**, 420.
WEBER, J.: (1973), Effect of concentration on laser threshold of organic dye laser. Z. Physik.
  258, 277.
WU, C.-Y., LOMBARDI, J. R.: (1973), Simultaneous two-frequency oscillation in a dye laser
  system. Opt. Commun. **7**, 233.
WU, C.-Y., LOMBARDI, J. R.: (1973), The effect of an electric field on the active medium
  in a dye laser. IEEE J. Quant. Electr. **9**, 26.

Subject Index

Introduced in January 1973

Applied Physics

Board of Editors	**A. Benninghoven,** Köln - **J. W. Goodman,** Stanford, Ca. **F. Kneubühl,** Zürich - **H. K. V. Lotsch,** Heidelberg **H. J. Queisser,** Stuttgart - **A. Seeger,** Stuttgart **K. Shimoda,** Tokyo - **T. Tamir,** Brooklyn, N.Y. **H. P. J. Wijn,** Eindhoven - **H. Wolter,** Marburg
Coverage	application-oriented experimental and theoretical physics:

Solid-State Physics   *Quantum Electronics*
Surface Physics   *Coherent Optics*
Infrared Physics   *Integrated Optics*
Microwave Acoustics   *Electrophysics*

Special Features	**rapid** publication (3-4 months) **no** page charges for **concise** reports
Languages	English with some German
Articles	review and/or tutorial papers original reports, and short cummunications abstracts of forthcoming papers
Manuscripts	to Springer-Verlag (Attn. H. Lotsch), P. O. Box 1780 D-69 Heidelberg 1, F. R. Germany

Distributor for North-America:
Springer-Verlag New York Inc., 175 Fifth Avenue, New York. N.Y. 100 10, USA

Springer-Verlag
Berlin Heidelberg New York
München London Paris Sydney Tokyo Wien

TOPICS IN APPLIED PHYSICS

Volume 2
edited by H. Walther will be about
Laser Spectroscopy
of Atoms and Molecules
with the following contributions:

H. Walther: General Survey on the Spectroscopic Applications of Lasers
T. W. Hänsch: High Resolution Spectroscopy Using Lasers
K. Shimoda: Double Resonance Spectroscopy of Molecules Using Lasers
E. D. Hinkley, F. A. Blum, K. W. Nill: Infrared Spectroscopy with Tunable Lasers

B. Decomps, M. Dumont, M. Ducloy: Linear and Nonlinear Phenomena in Laser Optical Pumping
S. P. Porto, J. M. Cherlow: Raman Spectroscopy of Gases
K. M. Evenson, R. Peterson: Light Frequency Measurements
J. L. Hall: Optical Frequency Standards and Precision Heterodyne Spectroscopy

SPRINGER TRACTS
IN MODERN PHYSICS

Editor: **G. Höhler**

Associate Editor: **E. A. Niekisch**

Editorial Board:
S. Flügge, J. Hamilton, F. Hund,
H. Lehmann, G. Leibfried, W. Paul

Volumes on Subjects
Relating to Lasers

Volume 66
30 figures. IV, 173 pages. 1973
Cloth DM 78,—; US $ 32.00
ISBN 3-540-06189-4

Quantum Statistics
in Optics and Solid-State Physics

R. Graham: Statistical Theory of Instabilities in Stationary Non-equilibrium Systems with Applications to Lasers and Nonlinear Optics.

F. Haake: Statistical Treatment of Open Systems by Generalized Master Equations.

Volume 70
to be published in February 1974
Quantum Statistics
G. S. Agarwal: Quantum Statistical Theories of Spontaneous Emission and their Relation to other Approaches.

In addition:
F. Schwabl, W. Thirring: Quantum Theory of Laser Radiation [in Vol. 36]

Springer-Verlag
Berlin Heidelberg New York

München Johannesburg
London New Delhi
Paris Rio de Janeiro
Sydney Tokyo Utrecht Wien